Public Urban Space, Gender and Segregation

The production of modern urban space in the Middle East is formed in the interplay between modernity, tradition and religion. Examining women in public spaces and patterns of interaction, this book argues that gendered spaces are far from a static physical–spatial division but rather produce a complex and dynamic dichotomy of men/public and women/private. Normative and ideologically-laden gender-segregated public spaces have been used as a tool for the Islamization of everyday life. In Iran, the most recent government efforts include women-only parks, purportedly designed and administered through women's contributions, to accommodate their needs and provide space for social interaction and activities. Combining research approaches from urban planning and the social sciences, this book offers an in-depth analysis of the morphological, perceptual, social, visual, functional and temporal dimensions of specifically women-only parks in Iran. It interprets power relations and the ways in which they are used to define and plan public and semi-public urban spaces.

This volume will be of interest to scholars and students in a wide range of academic disciplines including geography, urban studies, gender studies, political science, Middle Eastern studies, cultural studies, urban anthropology, urban planning, urban sociology, Iranian studies and Islamic studies.

Reza Arjmand is Senior Lecturer in the Department of Sociology, Lund University.

Routledge Studies in Human Geography

This series provides a forum for innovative, vibrant, and critical debate within Human Geography. Titles will reflect the wealth of research which is taking place in this diverse and ever-expanding field. Contributions will be drawn from the main sub-disciplines and from innovative areas of work which have no particular sub-disciplinary allegiances.

For a full list of titles in this series, please visit www.routledge.com/series/SE0514

Public Urban Space, Gender and Segregation

Women-only urban parks in Iran

Reza Arjmand

Routledge
Taylor & Francis Group

LONDON AND NEW YORK

First published 2017
by Routledge
2 Park Square, Milton Park, Abingdon, Oxon OX14 4RN

and by Routledge
711 Third Avenue, New York, NY 10017

First issued in paperback 2018

Routledge is an imprint of the Taylor & Francis Group, an informa business

British Library Cataloguing in Publication Data
A catalogue record for this book is available from the British Library

Library of Congress Cataloging in Publication Data
Names: Arjmand, Reza, author.
Title: Public urban space, gender and segregation: women-only urban parks in Iran / Reza Arjmand.
Description: Abingdon, Oxon; New York, NY: Routledge, 2017. | Series: Routledge studies in human geography | Includes bibliographical references and index.
Identifiers: LCCN 2016017518 | ISBN 9781472473370 (hardback) | ISBN 9781315603025 (ebook)
Subjects: LCSH: Public spaces–Iran. | Cities and towns–Religious aspects–Islam. | Urban parks–Social aspects–Iran. | Segregation–Religious aspects. | Muslim women–Iran–Social conditions. | Muslim women–Iran–Social life and customs. | Feminism–Religious aspects–Islam.
Classification: LCC HT147.I6 A75 2017 | DDC 307.760955–dc23
LC record available at https://lccn.loc.gov/2016017518

ISBN 13: 978-1-138-60111-6 (pbk)
ISBN 13: 978-1-4724-7337-0 (hbk)

Typeset in Times New Roman
by Sunrise Setting Ltd, Brixham, UK

Contents

Figures

Tables

Acknowledgments

This volume is a result of a project which was initiated during my tenure at the Center for Middle Eastern Studies of Lund University. Any scholarly endeavor is not possible without a long line of professional and personal support from many people of different capacities, at various stages and in many different ways, and this volume is certainly not an exception.

I owe a special word of gratitude to Masoumeh Mirsafa, from Politecnico di Milano. Without her generous efforts this project would have never been realized. Masoumeh has been engaged from the early stages of this project and not only conducted the fieldwork in women-only parks in Tehran and Rasht, but also helped with the data analysis and design of the figures.

I would like to thank and acknowledge the help of scholars and colleagues in the process of production of this volume: Reyhaneh Anjomshoaa for conducting the fieldwork in Niavaran and Bi'sat Parks in Tehran; Misagh Mottaghi for conducting the fieldwork in Isfahan, Shaqayeq Talebi, Kazem Ghalenoei, Zahra Molooki and Zahra Afraz for taking photos at different sites and parks.

I am grateful to Professor Leif Setnberg for his generous support of this project, and to all my colleagues at the Center for Middle Eastern Studies of Lund University. Special thanks to Professor Christa Salamandra at the City University of New York (CUNY) and Dr. Lisa Le Fevre of Teachers College, Columbia University for reading and critically commenting on different parts of the manuscript. I would like to extend my gratitude to Nadia Jahangiri for her generosity and support. I would also like to acknowledge the kind support and assistance of Maryam Ziari, Abuzar Majlesi Koopayi, Keyvan Ziaee and Ali Refahi at various stages of this project.

Transliteration guide

The transliteration method in this volume is a simplified version of the *International Journal of Middle Eastern Studies* transliteration style. However, familiar variant names follow the official spelling of the individuals or sites, even though they may not fully comply with IJMES system.

ء	'	س	s	ن	n
ب	b	ش	sh	ه	h
پ	p	ص	s	و	v or u
ت	t	ض	ż	ى	y
ث	s	ط	ṭ	ال	al or -l-
ج	j	ظ	z	ا	a
چ	ch	ع	'	◌	a
ح	h	غ	gh	◌	u
خ	kh	ف	f	◌	i
د	d	ق	q		
ذ	z	ک	k		
ر	r	گ	g		
ز	z	ل	l		
ژ	zh	م	m		

Introduction

What is space? Whatever theoretical perspective one adopts, one will have to accept that all space is constructed and that, consequently, the theoretical non-delimitation of the space being dealt with amounts to accepting a culturally prescribed and therefore ideological segmentation.

(Castells 1977: 234)

Sexuality and urban spaces as interplay between the body, space, social relations and power dynamics is a growing scholarly trajectory within the urban studies and social sciences. Space with its definitions, specific configurations, divisions and accessibility is the rendition of the social and political structures of society. It is, as Massey (1992: 81) puts it, "a moment in the intersection of configured social relations ... a complex web of relations of domination and subordination, of solidarity and cooperation." The politics of power are always sexual, even though space is a central mechanism of the erasure of sexuality (Grosz 1992: 246) and there is a constant interaction between the space and body, each informing and influencing the other. A spatial dichotomy based on a differentiation of female and male bodies and their functions in the urban space as well as the subsequent assignment of public and private realms has served as the foundation of spatial arrangement of modern cities. The city accurately embodies, among other things, the historic division of labor by gender within a normative structure. Gender relations are implicated in the conventional social and hierarchical arrangement of cities, where it is sanctioned that man should dominate space and that the house is the woman's assigned place (Lico 2001). The patriarchal structure of society permeates the body politics, in whatever form it may take, and justifies and naturalizes itself with reference to some form of hierarchical organization modeled on the (presumed and projected) structure of the body (Grosz 1992: 247).

Daphne Spain (1992) argues that the status differential between women and men creates specific urban spatial configurations linked to the patriarchal spatial institutions that reinforce the dominance of men. She maintains that the social system in place, through institutions of socialization, provides advantages to men that are denied to women. Hence, gendering of architecture is not straightforwardly visible since the values and ideologies architecture embodies claim universal status and are normally taken as gender-free. However, architecture as a system of representation is saturated with meanings and values which contribute to our sense of self and our culturally constructed identity.

(Lico 2001)

When spatial institutions are conceptualized and controlled by men, then the space within which they operate can be said to be biased in their favor and against women, making them effectively gendered spaces (Doan 2010). Hence, women's lives in urban spaces are shaped by the visible and invisible boundaries created by social structures. Miranne *et al.* (2000) notes that violence toward women is one of the mechanisms for perpetuating spatial dichotomy so that women who do transgress the spatial binary and enter public spaces must contend with an internalized fear of male violence. Women who enter male-dominated public spaces may be subject to a wide range of verbal and physical harassment for transgressing the established boundaries. In addition, other individuals whose identities reflect marginalized categories, such as race or sexual identities, also encounter this highly gendered spatial system and may feel especially constrained in the ways that they may express themselves in public spaces controlled by the dominant regime of power.

Space reflects the power symmetry of the social setting it resides in and is both controlling and confining of power and yet has the potential to disrupt these power relations (Duncan 1996: 128). Space is also an instrument of thought and action, which enacts the struggle over power between genders. Yet, it should be recognized that, as Lico (2001) notes, "space in itself is not inherently powerful." It is the politics of spatial usage that determine its power. A patriarchal framing of architectural spaces undeniably privileges masculinist power, in its representation of social order, hierarchical progression, polarities and stereotypical gender roles. An account on such spatial division would unveil the gender association both geometrically and symbolically across the traditional gender lines within society. This outlines the notion of sexual identity as the compulsory repetition of culturally prescribed codes or what Bourdieu refers to as *habitus* – the regulating systems of durable, transposable dispositions – which have become part of our unconscious. It is natural for us to think, feel and act according to a predefined set of images, languages and social practices, without inquiring as to the whys and hows of certain practices as we embody these gendered actions.

Bourdieu (1984: 170–2) also argues that habitus is the "structuring structures which organizes practices and the perception of practices" and reproduces and reinforces the dichotomous spatial division between female/male spaces. Hence, space is the "principal locus for objectification of generative scheme" which, with its bounding surfaces, enclosures, walls and levels, manipulates all bodily experiences. Hence as Colomina (1992) notes, "the relationships between sexuality and space hidden within everyday practices" and

securing space, in whatever form, is a political act: whether through invasion of territories, colonization, dispossession, appropriation, representation, the disciplining of knowledge or the purchase of real-estate. And the occupying of space is an assertion of power, and continual displacement is power's spatial effect.

(Taylor 1998: 130)

The power dynamics inform the space and shape the discourse around the spatial gender dichotomy. The nature of gender defined as a biological construct and its impact in shaping the public/private dichotomy has waked the criticism of scholars who questioned whether the gender binary continues to be a useful construct for looking at space. Judith Butler (1999: 12–13) argues that the socially constructed nature of gender makes it relevant to the culture which formulates the gender structure, and maintains that "bodies are understood as passive recipients of an inexorable cultural law when the relevant 'culture' that 'constructs' gender is understood in terms of such a law or set of laws." She challenges the traditional gender binary and argues that gender is not located just in people's physical bodies; rather, it is constructed through everyday performances of gender, which can challenge dichotomous conceptualizations and add fluidity to the range of possible gendered identities. While Butler, along with a group of other feminist scholars, endeavors to shift the binary and dichotomous structure of gender, others argue for recognition of gender-specific needs and accommodating the needs of women in urban spaces.

The persistent argument of the scholars and activists who call for accommodating the female-laden spatial division rest on the much-debated premise that no single public space can or should meet the needs of all users at all times and their variety is both necessary and valuable. In studying a particular public space or in theorizing about public spaces in general, it is important to consider the network or system of public spaces in which each space is embedded (Franck and Paxon 1989: 131). Henri Lefebvre (1991) suggests that the spatial patterns are not absolute but are shaped by the social and economic systems dominated by institutions and individuals who wield political power. In explaining gendered urban spaces, Rosaldo (1980) maintains that the division of space by gender is a product of social processes, not biological ones, and that viewing gender only in contrasting terms limits our knowledge and enforces a concept of women as different and apart from men instead of in relationships with men and with other women. While the notion of gendered space in terms of recognizing gender differences and understanding their needs and patterns of social interaction has been a concern in urban planning, it has turned into a central theme for feminist scholars across various disciplines as part of the endeavor to reinforce the traditional gender roles shaped by social institutions. Whereas construction of gender is conditioned by the patriarchal structure of society, feminist scholars criticizing the situation of women in urban spaces call for understanding the place of women in the public realm and study on the ways women and men experience urban spaces differently. In particular, scholars developed interest in analyzing the spatial expectations about women and their ability to move through urban spaces, to engage in labor outside the home, and to participate fully in the social and political system created and dominated by men (Doan 2010). The social construction of space is perceived as the rendition of the social structure. Sharistanian (1987) suggests that while the sexual asymmetry in the private and public realms seems to be universal, and can be seen in the ways in which it is organized, the actual activities and relationships women and men pursue, and the meanings they ascribe to them, differ among societies. Contested among feminist activists and scholars, a thoughtful distinction between private and public spaces continues to be employed as an analytical device across various disciplines. On the other hand, such premises fit well into the social structure among the wide range of societies, where

> men traditionally have exerted the greatest social and economic power and have influenced the spaces around them to meet their needs. Some locations benefit, and others are disadvantaged as a result of these dominant forces. In the same vein, groups without power are restricted from using the favored spaces, causing spatial inequality.
>
> (Doan 2010: 301)

Whereas the genesis of the gendered spaces can be traced back in the history formed by traditional, religious, cultural and normative values in various societies, the rise of industrialization and movement of production to places distant from the household is a hallmark of modern institutionalization of such normative values. During the Industrial Revolution, biological make-up and scriptural allusions were used to form the distinct ideology of "separate spaces" to accentuate the spatial gender division. While the public sphere was considered the world of men characterized by the production and wage labor of various economic, political and legal activities, the private sphere of women was expected to be the "proper space" for domestic life, child rearing, housekeeping and moral education. Such an institutionalized gender-based division rests on the argument that, since women and men are inherently different, the gender roles are natural and must form the basis of spatial division.

Such a premise is still employed in many societies as an ideal societal structure. Despite the fact that women have assumed a significant role in production and labor in the public realm, little has changed to accommodate and facilitate women's movement and interaction in the public space. In the West, the spaces of rapidly industrializing cities were considered unsafe for women; this perception led to the Victorian era division of space into public and private, which constrained women to the private space of the home and allowed men free rein to move through the public streets and seek out employment and entertainment in the city (McDowell 1983). However, the same spatial division persists in post-industrial cities where a larger proportion of the labor market is composed of women.

While there is a call for change in the public space to acknowledge the ever-increasing presence of women in the public sphere and to accommodate the spatial needs of women outside the private realm, in a parallel vein there has also been an scholarly endeavor to recognize that the opposition between "private" and "public" provides the basis of a structural framework necessary to identify and explore the place of female and male in psychological, cultural, social and economic aspects of human life. Hence, in the private/public dichotomy, the "private"

> refers to those minimal institutions and modes of activity that are organized immediately around one or more mothers and their children; "public" refers to activities, institutions, and forms of association that link, rank, organize or subsume particular mother-child groups. Though this opposition will be more or less salient in different social and ideological systems, it does provide a universal framework for conceptualizing the activities of the sexes. The opposition does not determine cultural stereotypes or asymmetries in the evaluation of the sexes, but rather underlies them, to support a very general (and, for women, often demeaning) identification of women with domestic life and of men with public life. These identifications, themselves neither necessary nor desirable, can all be tied to the role of women in child rearing; by examining their multiple ramifications, one can begin to understand the nature of female subordination and the ways it may be overcome.
>
> (Rosaldo 1980: 23)

Such dichotomous gendered structure of urban spaces still permeates urban studies and related disciplines. There is a call within the scholarly circles to deconstruct such "spatial binary which is used to legitimize the oppression of women" (Doan 2010) and suggest that such division of public and private might be best conceptualized using a kind of fractal analysis that breaks down the subcategories of space into geometric fragments. There is a need to move beyond this public–private duality and reconceptualize gendered space along a continuum (Gal 2005).

Attending to the needs of women in the public space, however, lends itself also to an entirely different discourse within the urban planning. This discourse is often fashioned around a religious belief or an ideology to legitimize the definition and division of the public space across the gender line. Mostly informed by normative and traditional values, such arguments are usually enforced by concerns over the safety and security of women in the public space along with the maintaining of the "sublime" position and dignified role of women to morally nurture the society. Islamic theocracy is not the sole instance of the practice and implementation of gender-segregated public spaces; however, is it perhaps the most contested one. Resting the argument on the notion of respecting the needs of women in public, the civil law based on *sharia* is utilized to institutionalize and legalize urban policies based on segregation of sexes. Whereas gender segregation has always existed as part of the culture and tradition in the Muslim context, the systematic effort to implement and institutionalize such divide is a rather recent phenomenon.

The gendered division of the urban spaces is undoubtedly an outcome of the socio-cultural processes. In the Muslim Middle East where religion has turned into an inseparable part of the tradition and culture, and gender is informed as the biological separation of sexes, gendered spaces follow the normative imposed by religion and tradition. Modernization, globalization and the steady pace of adoption of the Western lifestyle, however, have gradually affected the traditional structure of the urban spaces. While strict biological composition of sexes has imposed the gendered division of the spaces in many societies across the Islamic Middle East, there have always been endeavors to maintain the interactions between the two sexes in both public and private spaces. The Islamic city, hence, was formed "by the application of the Islamic legal system and usage to form a new and individualized urban fabric, in which private space was vouchsafed and public space was a feature of usage but without clear and marked boundaries" (Jayyusi *et al.* 2008: xviii).

In an attempt to define the distinct features of Muslim cities, Johansen (1979: 19–24) argues for demarcated separation between zones of economic activity and zones of domestic activity and residence. The strong centrality of urban organiza-tion determines the existence of two contrasted zones: a "public" zone occupying the city center and a "private" zone chiefly devoted to residence. In the "public" zones, those marked out by the presence of a broad avenue, a large market or an important mosque, responsibility fell on the political authorities. In the "private" zones, residential districts with cul-de-sacs, the people living in the neighboring houses were to answer for the consequences of any social misconduct there. Such spatial distinction, however, was the organic extension of a social structure with limited presence of women in public zones that assumed the private spaces of homes as the domain of women. Hence the separation of spaces in the Muslim city and "the way it was utilized, shaped and produced by different genders was not a simple case of dividing public–private geographies and assigning them to females and males, respectively" (Thys-enocak 2008). Abu-Lughod (1971), in her extensive analysis of gender interactions in the *hara* space in Cairo, challenges the notion of space dichotomy in the Muslim city and argues for the existence of a "third sphere," a semi-private or a tertiary realm, where gendered behavior is more fluid, the loyalties of family stretch beyond tribe or kin, and both women and men can move with greater ease.

Whereas the conceptualization of a third space is provided as a solution to bridge the gap between two separate gendered spaces, the notion of a third gender – as the mediator between the dichotomous spaces – is theorized and exhaustively discussed in religious literature of *hadith* and accommodated in some Muslim societies. The third gender known as *mukhannath* (eunuchs) – sexually ambiguous or biological males who identify as female or cross-dress to demonstrate the desire to change their biological gender – could freely move between two spaces separated across the gender line. With the recognition of such gender category, the space was not categorized as neutral or gender-free or accommodating a third gender. The third gender, however, functioned as mediator between two strictly divided spaces and could move freely between two separate female and male spaces, especially during ceremonies of vari-ous natures. There are numerous instances in Persian and Ottoman literature of *mukhannath* who were legitimized to move between two strict gender-specific spaces such as harems. Herdt (1994) argues that the third gender was recog-nized among Muslims to the extent that in the holy places of Islam *mukhannath* guardians were employed to move freely between strictly maintained female and male spaces. Reddy (2005) notes that the third-gender category of *hijra* was recognized in South Asia – including Muslim India and Pakistan – to play important public roles during religious ceremonies, among other occasions, to bridge the gendered spaces. Hence, one may argue that rather than creating a gender-neutral or mixed space, Muslims maintained the dichotomous gendered structure while finding a solution to mediate between them.

The gendered reproduction of the urban spaces in the Muslim context is the organic outcome of the power dynamics and reflection of the ideal gender roles as defined by the dominant ideologies within those societies. It

> has lain in the fact that the spatial structure embodies knowledge of social relations ... about the unconscious organizing principles for the description of society. Often a building is a concretization of these principles. They are expressions and realizations of these organizing principles in a domain that is more structured than the world outside the boundary.
>
> (Hillier and Hanson 1984: 184–5)

They also reflect ideals and realities about relationships between women and men within the family and in society. The space outside the home becomes the arena in which social relations are produced, while the space inside the home becomes that in which social relations are reproduced (*ibid*.: 257–61). Hence, urban planning and architecture set the conditions in defining the habitus of gender through distribution of bodies in space and delimiting and demarcating the interaction of female/male bodies in space. Architecture's enclosures and bounding surfaces reconsolidate cultural gender differences by monitoring the flow of people and the distribution of human subjects within the space (Lico 2001).

> You know how jealous the Muslim is of the integrity of his private life; you are familiar with the narrow streets, the façades without opening behind which hides the whole of life, the terraces upon which the life of the family spreads out and which must therefore remain sheltered from indiscreet looks.
>
> (Abu-Lughod 1980: 143)

Despite this, among Muslims the line between public and private is somehow blurred in many instances. The Iranian Revolution of 1979, which resulted in a modern theocratic administration based on religious law (*sharia*), advocated a homosocial culture in which women and men are expected to socialize separately. For many, this marked the beginning of an era of institutionalized gender segregation. The Iranian post-revolutionary government's policies demonstrated serious commitment to the separation of the sexes in the public domain. The ratification and implementation of a gender-segregation law (*siyasat-i tafkik-i jinsiyati*) in public spaces, from schools and universities to taxis, buses and sports centers, was an intriguing component in Iran's Islamization project. Gendered spaces were built or reconstructed around idealized roles of women and men in an Islamic society.

Against this background, the present volume aims to enhance understanding of interrelatedness of space, urban function and social interaction of women in a highly normative and segregated society. It further advances an understanding of the female body in public space as social polemics, a new front in the power struggle between two contradictory forces on two sides of the equilibrium. On one side of the forefront is the Islamic government's Islamization project which endeavors to engineer social relationships, manage the intimate and to normalize the gender-based segregation of public spaces; on the other side are women whose presence in public became a site of contestation for the Islamic government. Nonetheless, the extent and nature of women in public, and physical, functional and social measures by the Islamic government to inhibit such presence along with normative values, has left its imprint on the production and reproduction of both urban forms (physical and functional) and processes (social and psychological). Thus, the physical space "can be seen as 'sexed' and 'gendered,' not just the person who uses it" (Tonkiss 2005: 94). Women in such a context position themselves in relation to the center of power either identifying with it or positioning themselves against it. This results in a power dynamic which is based on "the articulation of Islam and patriarchy grounded in distinct material, social, political and cultural arrangements between genders" (Kandiyoti 1996: 24), and in turn leads to, among other outcomes, a distinct spatial reality.

Placement of the female body in the public space is the central notion of this volume. Women-only urban public parks as institutions, which are created and equipped based on a specific definition of gender relations, provide the possibility to study the female body in such a contested context as Iran. While the gender-segregation policy of the Iranian Islamic government has limited the presence of women in the public realm, gender-based public spaces are presented as an alternative. Hence, studying women in the urban parks (gender-mixed *vis-à-vis* women-only) provides the possibility of an in-depth understanding of such spaces and interaction of women in the public space.

A note on methodology and the disposition

The cross-disciplinary nature of this volume gains strength in part from an amalgamation of methods that respects both urban planning and social sciences. While the instruments of research largely follow the social sciences, the study's overall division adheres to urban planning, although with certain modifications. The methodological taxonomy for studying space, as scholars in such disciplines as architecture and urban design formulate it, suggests a multi-fold approach to exploring and scrutinizing various dimensions of a particular space. This approach includes looking at the morphological, perceptual, social, visual, functional and temporal dimensions. Such categorization may properly respond to projects within the field of urban planning, but becomes too detailed for studies of a cross-disciplinary nature. Hence, while remaining faithful to the framework suggested by urban planning, this volume introduces modifications to traditional categories so as to focus on three main dimensions for its analyses.

The first category includes the physical dimension of space, consisting of the morphological, perceptual and visual dimensions. The physical dimension has spatial characteristics stemming from a constructed space and includes environmental and aesthetic features that facilitate social life. This dimension is directly related to the intended function of a given space. Thus, the physical dimension addresses how a space's physical characteristics and construction can contribute to a place's ideal for both its social and functional dimensions. In this volume, physical dimension, as a category, examines "accessibility" of the parks, "legibility" both at the larger urban context and within the parks, the notion of "enclosure," and the "visual characteristics" of both women-only and gender-mixed parks.

The second category involves functional dimensions of the space, which also partially incorporates its temporal elements. Tonkiss (2005: 94) argues that a space's function "impacts the subjective identities and relations on three levels: on the meaning, the use and on the shape of urban spaces." Along with its physical and social aspects, the functional dimension of an urban space is pivotal in creating a successful and responsive public space. The functional dimension addresses how a place works and how people use it. Associated with various amenities, services and activities within a place, the functional dimension is a key contributor to user contentment. Hence, the main focus of this volume is on the three components of functional dimension for women-only and gender-mixed parks, demonstrating the way they contribute to a functioning public space. The first component, "mixed-use," addresses the variety of activities within a space and the way they influence the overall usage of the parks. The second is the concept of "adaptability" of parks in various social and temporal usages. The third component concerns issues of "management" and those of "surveillance" in the parks.

Finally, the third category focuses on social dimensions. Social dimensions discussed in this volume follow the methods of inquiry in urban planning to include a four-fold analysis of the parks as public spaces. "A place for all" aims to discuss relationships between people and space, and focuses on the egalitarian notion of public spaces and their availability to various social groups. The "vitality" of a public space is addressed in terms of the way parks affect and have been informed by various social activities. "Safety" and "security" within urban spaces pivots on the impact of safety as a key concept and the extent that safety enables the entry of women in public spaces. Ultimately, a "sense of place" explores the methods of (re-)appropriation and the notion of belonging in a public space. Working within these modified dimensions, spatial taxonomy becomes an analytical tool; however, neither categories nor dimensions should be considered mutually exclusive.

Eickelman (1981) argues that "the complexities of urban life [in the Middle East] pose particular challenges for analysis." He locates a major problem in studying urban spaces in the context of the Middle East as the "lack of communication across disciplinary boundaries." Sharing the concern, the present volume endeavors to take cross-disciplinary research on urban spaces to the next level by fusing urban planning, architecture and social studies. It draws on the fieldwork, including observation and semi-structured interviews with users of the parks, people and administrators, conducted in four women-only parks in three different cities (two in Tehran, one in Isfahan and one in Rasht) and two mixed parks in Tehran. Convenient and purposeful sampling method was administered to include the cases studied. Fieldwork was conducted intermittently from May 2014 through to May 2015. The data analyses suggest a pattern and confirm the analytical framework developed by scholars of urban design and composed of physical, functional and social dimensions. The author admits the fact that much of the discussion under the functional dimensions could safely be addressed under either of the former categories. However, drawing on urban planning and architecture literature, the functional dimensions were maintained as a category of its own, since it provides a tool to study the space in terms of its intended function. It could also reveal the gap between the designers' perception of a space and its ideal function and actual use of it, conditioned and constrained by social norms, a concept addressed and elaborated under "appropriated space." In Bourdieu's (1991: 113) terms,

> Appropriated space is one of the sites where power is consolidated and realized, and indeed in its surely most subtle form: the unperceived force of symbolic power. Architectonic spaces whose silent dictates are directly addressed to the body are undoubtedly among the most important components of the symbolism of power, precisely because of their invisibility. ... Social space is thus inscribed in the objective nature of spatial structures and in the subjective structures that partly emerge from the incorporation of these objectified structures. This applies all the more in so far as social space is predestined, so to speak, to be visualized in the form of spatial schemata, and the language usually used for this purpose is loaded with metaphors derived from the field of physical space.

While the idea that the relationship between space and social life is symbolic of power is fairly novel within social sciences, it is a rather complicated task to operationalize such correlations and argue for the perception of power through spatial structure. Hence, the debate over transformation of social space into appropriated physical space, and the role of power relationships and other normative values, needs further research that takes into account the specific configuration of cultural characteristics. The distribution and function of physical space is the rendition of social relations and power structures. However, when the structure of dominant power in the society is different from those of subaltern, the process of re-appropriation is likely to take another avenue; Michel de Certeau (1984) labels this as "tactics."[1] Tactics in the context of the present volume are, among other things, re-appropriation, redefinition and re-functionalization of the space, where the physical features of a space are compromised to redefine function. Hence the interplay between physical and social is to claim, re-appropriate and redefine the

> imaginary realm of ideologies of space and consider even these ideological and symbolic discourses as explicable with reference to the objective context of social relations. Power of spatial structures to represent something is an

irreducible feature of the symbolic and imaginary which itself produces social meaning that is not explained by an understanding of social processes.

(Prigge 2008: 47)

Space, thus, becomes a realm of everyday practice of power, with strategy (physical and functional design) as the instrument of the powerful and tactics (re-appropriation) as the tool of the powerless. This interplay between physical and social re-appropriation of the physical, functional and social, reflects the reality of social interactions within such spaces. "As physical space is defined by the mutual exteriority of its parts, so social space is defined by the mutual exclusion (or distinction) of the positions that constitute it, that is, as a juxtapositional structure of social positions" (Bourdieu 1999: 124).

Bourdieu (1999) also argues that social relations are interpreted and scribed in space. In the case of physical space, space creates divisions which reflect social distinctions. Physical space, thus, is the crystallized reproduction of the social. Space, like "field," becomes the venue where "power is asserted and exercised." The study of space, thus, becomes the study of social relations to the extent that it defines symbolic reality as more or less embedded in social context: "the naturalization of social processes conceals from us the process by which social reality is discursively constituted" (Laclau 1982: 17). The social context, with its interactions, relations and spatial discourse remains concealed, if it is not "theoretically reconstructed as a real object of discursive and symbolic practice and therewith recognized as a specific form of constituting the social" (Prigge 2008: 48). Using this notion, a space in its physical form and function can provide a narrative of history and a map of social power structure.

In the same vein, Foucault (1977) argues that architecture is "a diagram of a mechanism of power reduced to its ideal form" and assumes that in a disciplinary society visibility is a key instrument of control. Once again, the regulation of "biological and anatomical characteristics of the living human body" becomes the core of the power apparatus that seeks to regulate spatial division and management. Like "the hospital constitutes a means of intervention on the patient ... [and its] architecture ... must be the agent and instrument of cure" (Foucault 1979), similarly institutions employ spatial characteristics to attain the regulatory criteria.

Foucault's analysis of panopticon focused on an architectural form, rather than urban planning. However, his notion of visibility and its relation to the practice of power over the individual applies to the configuration of the urban space. Hence, the physical space becomes the embodiment of the mechanism of power across the society, a *dispositif* – a term that Foucault uses to refer to the various institutional, physical and administrative mechanisms and knowledge structures – consists of physical as well as social components. While the social dimension aims to enhance and maintain the exercise of power within a society, the physical structure translates such relations into a concrete form. "Foucault refuses to compare the architect to the doctor, the priest, the psychiatrist or the prison warden as professions through which power is exercised, because this power can only be applied via practices regardless of its physical environment" (Lambert 2013: 25). An architect, however, translates the structure and practice of power into a physical form and becomes the instrument of the power to embody the very "strategy" – ideological, normative or otherwise – into the form of a physical structure.

Similar to the field, a space is also a site in which the assertion and exercise of power is dominated by agents who possess adequate social capital. Agents, Bourdieu argues, struggle to appropriate space and these struggles take individual as well as collective trajectories. At the individual level, contestation occurs through families' spatial mobility and depends on class position. At the collective level, this struggle can be appropriated through public space policies. Bourdieu (1999) notes that the "political construction of space" used for public projects homogenizes space by bringing people who are socially and economically disadvantaged into physical proximity/contact.

In an endeavor to provide an in-depth understanding of the social settings and urban developments which gave rise to gender-specific urban parks in Iran, Chapter 1 explores the contested history of urban parks in Iran. It traces the debates over the extent to which they grew from indigenous Persian gardens. The chapter also follows the development of the *charbagh* model, which was extensively used in various parts of the greater Persia and subsequently across the Muslim world, as a visual articulation of the Celestial Garden. It reveals the sharp contrast between this form and the Western park model. Chapter 1 also analyzes the notion of *andaruni/biruni* as a spatial binary, and its impact on the social psychology of Iranians and the formulation and implementation of gendered spaces. Furthermore, post-revolutionary government policy on gendered spaces, its theoretical foundations as conceived through Islamic tenets, and the influence of ideological doctrines on gendered spaces are explored throughout. The chapter continues with an examination of women-only parks as a solution to the ever-increasing presence of Iranian women in the public realm, and addresses the social, legal, and demographic/geographical characteristics which form the context of those parks. It explores the social debates out of which the notion of women-only parks grew and traces the conception of women-only parks; a 2001 medical report attributed a series of epidemics among Iranian women to a lack of exposure to sunlight and a sedentary lifestyle. The chapter argues that despite the fact that medical ailments nurtured the discourse surrounding women-only parks, their construction served the Islamic government's overall gender-segregation policy. It offers a detailed account on the parks, their contextual and demographic characteristics, as well as the distinctive features of women-only parks compared to gender-mixed ones.

Built upon an analytical framework borrowed and altered for the purpose of the present study, Chapter 2 provides an in-depth account of the morphological and physical dimensions of the women-only and gender-mixed parks in Iran used as samples in this work. Faithful to a theoretical and conceptual model drawn from urban planning, the discussion is broken down into detailed analytical components. *Accessibility* of a public space, one of the main challenges in urban design, is studied in the light of both the technical characteristics found in macro (city-wide accessibility) and micro (in-park accessibility). *Accessibility* is also addressed from a social perspective and its contribution to social inclusion, not only in terms of socio-economic relations but also in terms of gender-specific challenges, such as with mothers with baby carriages and women with physical and/or mental disabilities. The chapter then turns to the *legibility* of the parks as public spaces with analyses at a macro level (city-wide) and micro level (in-park). A detailed account of the connectivity and legibility of each park form (both women-only and gender-mixed) is given, and an extensive analysis of both instances is provided throughout the chapter. Drawn from urban planning, the notion of *enclosure* illuminates the separation of different spaces for a variety of reasons. For women-only parks, enclosure is one of the most important design considerations. The enclosure of the parks and the normative-laden principles aimed at devoting a space for women in the midst of urban structures poses challenges for designers and policy makers who must incorporate the contrasting principle of *permeability*. Finally, the chapter addresses *visual attractiveness* as the aesthetic principle organizing the physical dimensions of parks as public spaces. The visual attractiveness of urban public parks reflects its function as a microclimate, as well as a venue for social interaction. In addition, the parks' natural structures like soft landscaping and greenery, and artificial ones such as the style of buildings, furniture, lighting, signage and symbolic landmarks, are explored and analyzed in this chapter.

Chapter 3 studies the functional dimensions of parks as spaces used by various groups of users for a variety of purposes. It contrasts the designers' intended function with the women's re-appropriations. Through a discussion of *mixed-use* – a key concept in designing urban public spaces – the chapter examines the variations in use and adaptation or appropriation of parks as public spaces. The possibility of serving the needs and demands of various groups of users from different socio-economic backgrounds and normative affiliations and interests is addressed and discussed. While *mixed-use* addresses individual users' needs, *adaptability* refers to a space's capacity to accommodate social, economic and technological changes. The notion of *adaptability* as a functional feature relates to a division between indoor and outdoor spaces within the parks. The chapter also provides an elaborated account of the impact of adaptability on women's use of space. Finally, Chapter 3 discusses surveillance and methods of management. Departing from Foucauldian governmentality and biopolitics, the notion of control and power dynamics is explored through of the distinction between *hard (active)* and *soft (passive)* surveillance. While the hard method is implemented by the utilization of moral police, security officers, surveillance cameras and searches (bodily and belongings), the soft method uses symbolic restrictions and exclusion from enjoying certain opportunities or activities.

Urban parks are regarded as venues for social interaction. Chapter 4 endeavors to address the relationship between people and space, analyzing parks as sites of network building (among other things). While posing certain limitations on social interaction, women-only parks open possibilities for new forms of networking.

This becomes critical given the lack of state and public institutions to create and enhance social networks. Mothers with infants and grandmothers with their grand-children are examples of such networks that, in the absence of any public policy or institutional support, create informal networks that use parks for their regular visits. Specific to the women-only parks is the network of homosexual women who exploit the parks for visits. They consider the parks to be "safe havens" for their meetings, which they conduct free from the forms of harassment they would experience in mixed places, from random men or the moral police. Again, based on frameworks drawn from urban planning, the specific social dimensions are studied through a number of detailed concepts. *A place for all* argues for social inclusiveness of the parks as public spaces. The chapter addresses the challenges that the exclusiveness of women-only parks poses to the notion of a public space as a space for all. *Vitality* as a result of human interactions is another point that the chapter addresses. This includes the flow of people (pedestrians and others) and the possibilities for interactions and events. Chapter 4 concludes with a dis-cussion of *safety* and *security* as the central factors affecting people's presence, interactions and use of space. The notion of *sense of place* and *defensible space* addressed, as *feeling unsafe* is compared and discussed in relation to a lack of safety and security.

Women-only parks were received differently by various groups of women in Iran. Whereas the idea of a gender-specific park faced resentment from some groups of women, especially women's rights activists and feminists, others found it – however restricted and controlled – an opportunity to experience a public space without *hijab*. Building on interviews conducted with women who favor women-only parks and actively make decisions to use them, the first part of Chapter 5 reflects on the arguments of various proponents of women-only parks. The argu-ments of the opponents to women-only parks are presented in the second part of the chapter. In the same vein, the ideas of women who actively denounce the idea of women-only parks and perceive them as an extension of the gender-segregation policy affecting aspects of Iranian life are presented and discussed. Using critical ethnography, this chapter (like Chapter 4) reflects in-group variation by providing a diverse array of opinions and ideas.

The final chapter concludes the discussion on gendered public parks in Iran by providing a summative discussion on women in public spaces and compar-ing the women in the women-only parks and gender-mixed parks. The impact of each space on the production and reproduction of social interactions of women, as well as on the polemics around the role of women in public space, concludes this volume.

Note

1 de Certeau (1984) links "strategies" with institutions and structures of power which are the "producers," while individuals are "consumers" acting in environments defined by strategies by using "tactics." In the influential chapter "Walking in the City," de Certeau asserts that "the city" is generated by the strategies of governments, corpora-tions and other institutional bodies who produce things like maps that describe the

city as a unified whole. de Certeau uses the vantage from the World Trade Center in New York to illustrate the idea of a synoptic, unified view. By contrast, the walker at street level moves in ways that are tactical and never fully determined by the plans of organizing bodies, taking shortcuts in spite of the strategic grid of the streets. This concretely illustrates de Certeau's argument that everyday life works by a process of encroaching on the territory of others, using the rules and products that already exist in culture in a way that is influenced, but never wholly determined, by those rules and products.

1 Urban parks in Iran

> The heavenly breeze comes to this estate,
> I sit with the wine and a lovely mate.
> Why can't the beggar play the king's role?
> The sky is the dome, the earth is my state.
> The green grass feels like Paradise,
> Why would I trade this for the Garden gate?
>
> *(Hafiz)*[1]

Urban green spaces for recreational purposes appeared long before urban public parks. La Alameda, a garden square in Seville (built in 1574) is suggested as the first known public garden with a function similar to modern urban public parks in Europe (Albardonedo Freire, 2002: 194) which promoted the notion of the modern urban parks. Nevertheless, there have been institutions in other parts of the world, however different in forms and structures, to fulfill similar functions. Among scholars, there has been disagreement as to whether the modern parks in Iran are the result of the development of the indigenous Persian gardens or an idea borrowed from the West in the process of modernization. The Persian garden (*Bagh-i Irani*),[2] a green man-made space and an urban microclimate with its distinct elements, is presented by some scholars as the Persian vernacular version of the modern park. Using terms such as *bustan* or *pardis* in lieu of park after the Islamic Revolution has also strengthened this assumption. Although using certain terminology might appear as an endeavor to use the familiar terms over loanwords, in such an ideologically-laden context as Iran it also defines the ideal function attributed to the place. A *pardis* (in all its forms, including *bagh* and *bustan*) is perceived as a visual articulation of the Celestial Garden or divine paradise on Earth. Persian gardens, imperial or vernacular, private or public have been an integral part of Iranian architecture and urban planning. Instances of such gardens were referred to in scholarly literature and excavated in Pasargadae, Persepolis, Susa, and other ancient Persian sites (Pinder-Wilson 1976: 83).

This chapter provides a brief historical account on the evolution of the urban green spaces in Iran and discusses whether the modern urban parks in Iran are the modernized and appropriated forms of Persian gardens (*bagh*). The chapter also explores the distinct quadripartite (*charbagh*) model, which was initiated in

Persia and subsequently adopted across the Muslim world for its "highly important dimension of symbolic meaning as earthly reflections of the paradise that awaited the faithful" (Ruggles 2014). It also addresses the cultural roots of gendered spaces in Iran and how such dichotomous notions as *andaruni/biruni* laid the foundation for gendered public spaces in Iran. The social polemics around the women-only parks in Iran and the Iranian government's endeavors to implement the segregation policy across various segments of the society will also be discussed in detail here.

Baghs: an Iranian indigenous form of park? A historical account

The first gardens in Iran were established during the Achaemenids' reign (550–330 BCE) as part of their interest in the development of an indigenous horticulture and agricultural method. They "encouraged the efforts … toward innovative practices in agronomy, arboriculture, and irrigation" (Fakour 2000). Hence, gardens were created not solely for horticultural and sensual pleasure purposes but also incorporated political, philosophical and religious symbolism as an indication of authority, fertility, and legitimacy (Eliade 1961: 59–72; Stronach 1976). Fakour (2000) argues that, "what made gardens special during the Achaemenid reign was that for the first time the garden became not only an integral part of the architecture, but was also the focus of it." Henceforth gardens were an integral part of Persian culture. Achaemenids also set a standard for *charbagh* as a model of Persian gardens, the earliest of which are located at Pasargadae, the royal park residence of Cyrus the Great (*c.* 559–530 BCE). The royal palaces at Pasargadae were conceived and constructed as a series of palaces and pavilions placed among geometrically designed gardens, parterres, and meticulously hewn and dressed stone watercourses, set in a large formal park containing various flora and fauna (Stronach 1976: 107–12).

One other innovation, which is widely acknowledged as a pivotal component contributing to the creation of urban green spaces across the Persian plateau, is *qanat* or *kariz* (underground water channels) technology. Exploited as early as the first millennium BC in ancient Persia as a solution to the scarcity of water in an otherwise arid area, *qanat* left its imprint on Persian architecture and urban design for which the use of natural slopes, however minor, turned into an advantage to be exploited creatively. This principle is extensively used in various urban structures, not least gardens, private as well as public, and transferred as a distinctive element in Persian vernacular urban design.[3] *Qanats*, which through a web of channels carried the water over long distances and from deep underground up to the surface, created yet another possibility: the open water channel (*juy*), which became a standard feature of the traditional Persian public spaces including streets and gardens alike.

It was, however, during the Sassanid era (224–651 CE) when the pattern of the Persian city with private gardens and larger public *pardises* as inseparable components of Persian urban design were completed. The Persian *pardises* were inspired by the city's design and in turn informed and influenced the urban structure of the

Persian cities. They turned into such a pivotal element in Persian urban design and architecture that, in many instances, the urban fabric was designed and built according to or to accommodate a garden or *pardis*.

Traditionally, Persian *baghs* were private spaces where "a great wealth [was] required to finance the acquisition, development and maintenance of a formal garden, especially in the Persian arid landscape and made this type of holding a symbol of power and prosperity" (Fakour 2000). Little is known about the actual form of the Persian garden before the Islamic period other than its *charbagh* shape, but its existence at that time and its importance as both a symbol of power and resource for pleasure is widely acknowledged (Pinder-Wilson 1985: 71–3). While the larger body of literature meticulously focuses on various aspects of Persian gardens from urban and architectural outlooks, there is a scarcity of sources on their social aspects and only traces of their social functions can be extracted from travelogues and other scattered sources on gardens.

Arguably the earliest known instance of Persian public gardens dates back to the Buyid reign when 'Azod-ul-Dawla (949–983) constructed a royal suburb outside the city of Shiraz, called *Kard-i Fana Khosrow*, which contained an extensive commercial district as well as large gardens and palaces (Muqaddasi 2014: 430–31; Golombek 2000). Later, the Ghaznavids (998–1186) developed extensive gardens in the city of Ghazneh with which the public use of gardens took a turn. Sultan Mahmoud's (d. 1030) burial site became the celebrated *Bagh-i Piruzi* (the Garden of Victory) with the intention of attracting more people to visit the mausoleum of the sultan. Upon the success of the Memorial Garden of Victory as a grandeur reminder of the power of the Sultan and his successors; Mahmoud's heir Sultan Mas'ud I (1030–1040) founded numerous *bagh*s in or near the major towns of Balkh, Herat, Bost, Nishaboor and Ghazneh and moved about between them occasionally, spending little time in any one place. There is no documented evidence to suggest that those gardens were open to ordinary members of the public; however, numerous sources (Allen 1988; Bosworth 2007; Fakour 2000; Bennison and Gascoigne 2007: 155) note a number of installations including a *kushk* (chateau), *a maidan* (large square), a palace with a grand public reception annex and a bazaar, which could implicitly mean that the *baghs* – partially or entirely – were open to the general public.

The Safavid era (1501–1722) was probably a turning point in redefinition of the Persian garden, from a private closed to an open public space. It was during this era that the Persian *charbagh*, the classic quadrilateral model divided into four smaller parts by walkways or flowing water, was re-appropriated and used as a public urban space. The highly structured geometrical layout was used, among other spaces, to define and build the famous Chahar Bagh in the Safavid capital city of Isfahan. Sometime during the fifteenth century, Chahar Bagh, which has survived to this day and still is used as a pedestrian boulevard, was incepted as a public park. A European visitor who visited Chahar Bagh in 1666 describes it in his journal, which provides a portrait of its physical form as a public *bagh* with some functions similar to those of the modern public parks.

We pass through the most charming parts of the Chahar Bagh, taking a course over its alleles of unequal plain trees, stretching their broad canopies over our heads, their shade being rendered yet more delightful by the canals, reservoirs, fountains, which cool the air, and reflected the flickering light through their branches. Thickets of roses and jasmines, with clustering parterres of poppies, and other flowers embank the ground; while the deep-green shadows from the trees, the perfume, the freshness, the soft gurgling of the waters, and the gentle rustle of the breeze combining with pale golden rays of the declining sun, altogether form an evening scene, as tranquilizing as it was beautiful.

(Chardin 1811: 118)

Despite the wealth of historical materials on Chahar Bagh of Isfahan or other (semi-) public *baghs*, there is little study to shed light on human interactions and the everyday life of people in the public space. Numerous sources point out the centrality of the Chahar Bagh as a hub to connect a wide variety of social and urban functions and activities. Hence, one may assume a central social role for Chahar Bagh as not only a place to socialize, meet and enjoy the vicinity of nature but also its utilization as a connection route to other urban functions. All significant urban functions of an Islamic city including bazaars, *madrasas*, mosques, coffeehouses, the royal polo yard and royal palaces were built around the rectangular formed Chahar Bagh. Despite the scarcity of materials on human and social dimensions of *baghs* as many studies were concerned with the forms and styles of such garden rather than social dimensions, they render a portrait of the social functions of such Persian public gardens as Chahar Bagh of Isfahan.

Despite the abundance of historical facts on the gardens – both private and public – in Iran, certain crucial questions persist to this day: Was a Persian garden (*bagh, pardis*) similar to a modern park in form and division of space? Did those gardens fulfill similar social functions to the modern parks? And, whether the design of modern parks in Iran incorporated or was informed by indigenous elements of the Persian garden? In an effort to provide answers, this chapter endeavors to explore the phase of transition from gardens to parks in Iran. Koopayi *et al.* (2013: 4), in the study of Amin-ul-Dawleh Park (established in 1891) as the first modern park in Tehran (or perhaps Iran), provides a detailed account on the similarities and differences of Persian gardens and the first modern urban park, and notes that Tehran's administration of the time was well aware of the differences between these two institutions and hence they borrowed and introduced the term park into Persian lexicon to mark the difference between the two public spaces. With the rapid modernization and urbanization based on Western urban design, Tehran's French mayor Alexander Buhler adopted the city of Paris's urban plan as a new urban development plan for the city of Tehran. The plan marked certain spaces as *baghs*, others as meadows and designated three spaces specifically as parks. While there is a movement to recognize a revival of the indigenous Persian institutions and strengthen the cultural elements of the parks to resemble Persian gardens, there is a consensus among urban scholars that parks are not extensions of the Persian garden and lack both the Persian indigenous components and vernacular features of the local architecture, as they are meant to fulfill a different social function. As Sultanzadeh (2013: 95) argues, introducing some Persian elements into modern parks would not turn a park into a garden. Parks are modern institutions, established to respond the needs of the modern man in a modern urban setting.

The Qajar reign (1785–1925) is regarded as the era of modernization in which the European lifestyle was introduced and adopted both by the administration and by intellectuals. Also, modern lifestyle became the criteria to distinguish between different social classes. As a result, an interesting development occurred in architecture and introduced a series of changes into the otherwise vernacular urban design. While the public buildings, or public parts of the private houses (*biruni*) followed the Western so-called extrovert architecture, private parts (*andaruni*) were still designed and built using the same introverted traditional model (Etemad-ul-Saltaneh 1999: 86). Sultanzadeh (2013) suggest that, using such dichotomy, the owner while acting modern in public was marking the dominance of traditional values in family life. Despite the persistence of Iranians in maintaining their traditional values, the slow pace of westernization continued to progress and to affect Iranian lifestyle and urban structure. With the return of waves of West-educated Iranian architects, the process hastened and changed the face of the Iranian urban life forever.

Parks were among the public spaces that were introduced in the process of modernization of the urban lifestyle in Iran. The main physical difference between the traditional Persian gardens and parks was the use of curved lines and organic design in the parks instead of the straight lines of the traditional *charbagh* model. Using landmarks such as statues and sculptures is considered a demarking component of the modern parks, since sculpting was not such popular art genre among Persians after Islam. It is argued that, Iranian traditional architecture failed to introduce innovative methods to adapt to and accommodate the rapid modernization of the lifestyle and lagged behind in creating such public spaces as Persian parks. Along the way, and as a result of such failure and the zeal of the people to be modern, *baghs* were replaced by parks, which in many cases were copied directly from Western countries. Thereafter, there were some sporadic efforts to introduce elements of Persian *baghs* into modern parks and bestow upon the modern parks a Persian identity.

Henceforth, Nasir al-Din Shah's era (1831–1896) is also regarded the first phase in the introduction of Western lifestyle in Iran, through the initiation of modern institutions, commercial and cultural exchanges with Europe and opening embassies to facilitate travel as well as the multiple visits of the Shah and people at policy and practice levels to European countries. In his first travel diaries to England and France, the Shah gives an account of Hyde Park and distinguishes between Iranian gardens (*baghs*) and Western parks. For him, unlike a *bagh* a park was not solely a green space to serve the purpose of tranquilizing urban inhabitants. Upon his return to Iran, he coins *bagh-i 'amme* (public garden) to distinguish between traditional Persian gardens and reserves the term "park" for public spaces of certain social tasks.

Setting standards for green spaces and urban parks in Iran

Urban parks as green spaces within the city are created and maintained with the intention to contribute to the quality of life of people in an urban setting. Among many acknowledged benefits, urban parks increase the quality of air, help to decrease the heat island effect and positively contribute to promote the health of the urban inhabitants. Such features turn urban parks into significant components of an urban context and an indicator to gage the quality of urban life. This has resulted in a series of regulations and standards, both at national and international levels, to suggest (minimum) required standards to meet the needs of an individual in having a green area in an urban surrounding and improving the health of urban residents.

The World Health Organization (WHO) and the United Nations Food and Agriculture Organization (FAO) suggest a minimum availability of 9 square meters of green open space per city dweller as an international benchmark for the minimum standard (Konijnendijk 2005; Kuchelmeister 1998). Having the flexibility to adapt to the realities of certain countries, the minimum standard is usually modified by various countries. The Iranian Ministry of Housing and Urban Planning advised a minimum standard of 7 to 12 square meters per capita as the Iranian standard. The figure, however, varies from one geographical area to the next and is interpreted and implemented differently in various climatic zones across the country.

From an urban planning perspective, urban population and the total area of green spaces within the city are two main parameters in calculating the urban green space per capita. Moreover, the geographical distribution of such spaces also plays an important role in the assessment process. Such characteristic becomes even more decisive in the context of larger cities where parts of a city may suffer from a scarcity of green space, and most green spaces are concentrated in other parts. This also suggests that in order to keep up with the growth of the population, the green space in urban areas must constantly be expanded in the direction of the urban expansion to keep pace with the population increase.

Meeting this demand, article 55 of Iranian Municipality Act stipulates the erection, development and maintenance of public green spaces among the responsibilities of the municipalities. Municipalities are in charge of different types of green spaces within their catchment area. With the rapid expansion of Iranian cities and the need to maintain the pace of expansion of green spaces, the Organization for Urban Parks and Green Spaces was incepted as a responsible body for the management, development, expansion and maintenance of urban parks, green spaces, squares and recreational centers. It also leads and conducts scientific research on green spaces and planning for future development across the city.

In Tehran, the Department of Gardens, a subsidiary organization of the municipality established in 1960, was the first of its kind in the country, responsible for managing and maintaining the urban green spaces of the capital city. In 1963 it was renamed "Parks Organization" to promote the role and enhance the importance of green spaces in urban life. Through the growth and development of the urban green spaces inside the city, the organization was assigned to assume greater responsibilities. This resulted in renaming the organization in 1990 to Tehran's Parks and Green Space Organization (TPGSO). Ever since 1990, TPGSO has been responsible for maintenance and development of the green spaces in Tehran. Management, development, maintenance, and supervision of public open spaces, recreation centers, tourist resorts, parks and urban greenbelts are among the main tasks of TPGSO (Manucipality of Tehran 2015).

The establishment of TPGSO served as a model for other cities across the country to emulate in order to plan and develop their own urban green spaces. The Urban Parks and Green Space Organization in Isfahan – the third largest city in Iran and one of the cases studied throughout this volume – was established in 1987 to "monitor, develop and improve the green spaces within the city" (Isfahan Municipality's Digital Portal 2015). With numerous city-wide branches covering different districts, the Urban Parks and Green Space Organization has been an active body in developing green spaces across the city (Isfahan Municipality's Digital Portal 2015). By the end of 2010, the urban green space per capita in Isfahan had reached 24.2 square meters and it was envisioned that the green spaces of the city would be developed to reach the target of 30 square meters per capita (Isfahan Municipality 2015).

In Rasht – a medium sized city by Iranian standards and another city studied in this volume – the Organization for Urban Parks and Green Spaces is one of the recently established bodies subsidiary to the Rasht Municipality and has three branches across the city to manage the green spaces. As part of endeavors to introduce a culture for using the urban green spaces, the Rasht Municipality aims to enhance green spaces within the city and encourages inhabitants to use such spaces. According to the Rasht Comprehensive Plan 2008, the total area of green spaces in Rasht is about 245 hectares, which provides 4.5 square meters per capita, low indeed compared to the adopted standards. Given the geographical, topographic and climatic characteristics of the city, there is a great potential for the development of its green spaces. Hence, the Urban Parks and Green Space Organization of Rasht is planning to meet the vision of reaching the optimum standard of 30 square meters per capita.

Women-only parks: from idea to implementation

The rise of a gender distinctive urban culture in Iran dates back to the Sassanid Empire (224–651 CE) when the built environment reinforced sexual differences, articulated assumptions about gender and symbolically embodied gender identities (Rizvi 2000; Karimi 2003). Rizvi (2000) argues that Safavid women, through their patronage to various public spaces,[4] mostly shrines, claimed their authority over society and made themselves visible in the public realm. The royal women emerge as autonomous actors and exploit patronage to not only emphasize the active role of women in society but also their share of an otherwise patriarchal structure of the government. Timurid princess Gawhar Shad (d. 1457) and the way she exploited codes of piety and politics of the fifteenth century to reinterpret and use the religious expression for her activities, served as role model of a woman who was highly engaged in social issues, not least through building madrasa for male clergies.

Despite the vibrant and active participation of highly educated and socially active women affiliated to the royal family and elite classes in social arenas, during the Safavids and later the Qajar dynasties women were excluded from public spaces and women's presence and mobility in urban public spaces were restricted. Karimi (2003: 28) argues that the seclusion during Qajar was often regarded as a sign of honor, seen more stringent in wealthy neighborhoods where houses were surrounded by tall walls with no window opening to the streets.

The Pahlavis (1925–1979), however, introduced Western values as part of the modernization effort in Iran which introduced gender-neutral public spaces in the country. Although unveiling became part of modernization discourse in Iran and to be modern was emphasized by unveiling, a conscious resistance against the Pahlavis' policies on women was demonstrated by committed religious factions. The female body and its veiling and unveiling became the domain of polemics and struggle in social scenes in Iran. Despite all, the Pahlavis' efforts contributed greatly to the visibility of women in the social arena in modern Iran.

With the Islamic Revolution of 1979, veiling turned into political symbols to emphasize, "the difference from the Western world" (Göle 1996) and the pre-revolutionary values alike. Veiled bodies of women in public were to "consolidate the image of the Islamic Revolution" (Amir-Ebrahimi 2006). As part of de-secularization and revival of religious values in Iran, the process of re-veiling was imposed by the Islamic government and veil (*hijab*) became a precondition for the social and public presence. On the other hand, due to the strong influence of '*urf*,[5] even when there were no legal hindrances to women's public presence, the presence of women in public spaces such as parks was not welcomed by religious- and traditional-minded people. Hence, despite the fact that the public spaces were not engendered or defined as such, an unwritten, informal law reigned to divide public spaces across the gender lines. The presence of women in such spaces, thus, became the token of modernization and indicated the extent of westernization, and marked a self-claimed affiliation to either of those groups. After the revolution, the presence of women in public space was conditioned to abide to the normative criteria by the Iranian Islamic government. The veil (*hijab*), though perceived as a limiting factor, became the tool to access social mobility and be active in the public space.

The gendered spaces, albeit institutionalized after the Islamic Revolution in Iran, has never been an alien concept in Persian traditional urban design and architecture. In many public (or even private) spaces, as a result of observing rules of either religion or tradition, separate spaces were designated to women and men. Iranian houses were designed to accommodate this strictly maintained structure and divided into an interior space (*andaruni*) and an exterior (*biruni*).

However, this physical division at a micro level also resulted in a conceptual division in society at large. Symbolically, the gender division of a private house was a microcosm of the gender divide in society and started from behind the entrance door before even entering the house. Two different doorknockers (one for females and one for males) with different sounds were mounted on the entrance. Using the right knocker, the visitor to the house announces his/her presence to the respective inhabitants. While *andaruni* is regarded as a private space in which no dress code is observed and female members of the family (*namus*)[6] mix and mingle freely with their kin (*maharim*) without observing *hijab*, *biruni* is an extension of the public space at home. Iranian architecture is a rendition of the social life in society. The division of space into private and public with two separate entrances that conceptually divide people into those who are entitled (*khodi*) and those who are not has reached a much wider context than Iranian homes. The *andaruni/biruni* division created two parallel worlds that co-existed alongside each other.

> Not only was the house divided into two but everything inside the *andaruni* and *biruni* was divided into twos. Each of the two sections was built around a garden, each garden was clearly divided into two. As you entered there was a right-hand side and a left-hand side, each mirroring the other. ... the total isolation of the more public "outside" from the "inside" part of the house was one rule of division in two that no one dared violate. There was no opening, not even a window that joined the *andaruni* and the *biruni*. ... For those permitted into the *andaruni* everything was to some degree public, and you had privacy only insofar as you were able to feel private in your mind.
>
> (Mottahedeh 2004: 26–9)

Biruni, which was the public or male quarters of the Persian houses, was used also for the conduct of business, male religious ceremonies and parties (births, circumcisions, weddings, return from pilgrimages and funerals) for men. It was less furnished than the *andaruni* or women's quarters and had separate (and smaller) courtyards planted with fruit trees, shrubs and flowers, and set with pools and fountains. It contained a guest room for visiting male family members and a small pantry (*abdarkhaneh*). The interaction between the two spaces took place through a messenger – male members of the family under the age of puberty or boys hired for that purpose, but never girls. Once the boys reached adulthood, they were dismissed or else taken on as full-time servants in a different capacity. When there was a need to send a message or for food, the messenger would be sent to the *andaruni*, where the kitchen, pantry, and storerooms were located (Djamalzadeh 1985).

The *andaruni* was considered a space of domestic intimacy. Strong, high walls presented an outside face to private residences, with no windows for public exposure,[7] but through a door leading into a narrow covered passageway (*dalan*) one emerged into the courtyard. *Dalan* also indicated that male outsiders should proceed no further. The entrance to the *andarun* was sometimes indicated by no more than a thick canvas curtain, although in the houses of the truly affluent, eunuchs (*mukhannath*) were posted as guards.

In cases where the landlord had multiple wives, whether permanent or temporary (*mut'a*), there were multiple *andarunis,* one designated to each wife. In such cases the term *haram* or *haramsara* is used respectively. While observing norms of *sharia* and tradition was among the prime reasons for such division, among religious minorities such as the Zoroastrians, according to Boyce (2000), concern for defense and security sometimes led to analogous closed arrangements.

The idea in providing such a detailed account about this binary spatial opposition in Iranian culture is to argue that through a process similar to "generification"[8] such spatial dichotomy has been re-appropriated into the public sphere. With the dominance of the Islamic system and "establishment of the Islamic Law in different aspects of daily life, new codes of appearances and behavior have changed Iran to a 'metropolitan *andaruni*' where everything was defined by the ethical codes of conduct and appearance" (Amir-Ebrahimi 2006). Some instances of using *andaruni/biruni* taxonomy could be seen in the public realm in Iran today. The Laleh Hotel in Yazd, for instance, reserves a space for exclusive use by single female guests in its *andaruni* (private) section, which formerly served as the women's quarters in the residence (Lahiji 2006 as cited in Karimi 2003).

As argued by Amir-Ebrahimi (2006: 456–7), during the 1980s, the absence of public spaces for women in Iran resulted in their invisibility in the public social arena. Underlying Iranian female space in society is a homosocial culture where women and men generally socialize separately which causes space to be gendered around the idealized roles of women and men (Kakar and Bauer 2003: 536). Historically, the various forms of gendered spaces were introduced and utilized in Iran. In many instances, rather than creating a women-only space, the arrangement was to use the space in turn. Abedi and Fischer (2006: 322) describe one of such instances in an Iranian village, where "the bathhouse was used both for cleanliness and ritual purity. Menfolk used it before dawn, womenfolk afterwards." Kakar and Bauer (2003: 537) provide various examples of such division in rural areas where the female space extends beyond the private sphere to relate the tasks or social functions with what an ideal woman should do. Women are expected to wash and cook and so certain times of the day and/or access points to water are female spaces. Thus, particular wells may be designated female spaces, or if shared with men, certain times may be designated as women-only (usually early morning and late afternoon). The solution was used in the larger-scale public realm after the Islamic Revolution in Iran along with other methods.

The revolution opened a new chapter in Iranian women's public life. Veiling became an important part of the new social and political discourse. In 1936, some four decades before the revolution, Reza Shah (1878–1944) abolished veiling of women in public to inculcate a secular and Western lifestyle. In a similar vein, theocratic administration imposed re-veiling as an emblem of a new era based on Islamization. It revered veiled women as the ideal and devalued secular women as Westoxicated (*gharbzadih*), monarchical (*taghuti*) and indecent. Re-veiling became more than a shift in the dress code, it served as a centerpiece of the Islamization policy. Sedghi (2007: 200–2) argues that a process of renewal, re-appropriation and extension of the old concept of *namus* introduced a new interpretation of the role of women in the public domain. Communicated through wall graffiti, leaflets, word of mouth, the media and the new officials, an old Iranian cliché returned to the foreground: "the Woman Represents the Chastity of the Society" (*zan namus-i jami'i ast*), which in turn led Ayatollah Khomeini and the revolutionary government to enact a series of incentives, policies and practices. Women were seen to bear a heavy responsibility for the moral health and "therefore the political fate of the country [and] … women's sexuality is accorded tremendous power over men and provides the basis for all the arguments for segregation and veiling of women" (Najmabadi 1991: 67).

The polemics around forced veiling was escalated when the gender-segregation policy initiated in public beaches and sports facilities and spread to other spaces and activities across the country. Ayatollah Khomeini declared it a religious obligation (*vajib*) on March 29, 1979. Only three days later, Khomeini concluded the much-heated debate about veiling and asserted veiling (*hijab*) compulsory for state employees. Despite women's protests, re-veiling became a legal obligation for all women across all sectors in public. Hence, the formal process of compulsory veiling took place in two stages. The first stage consisted of targeting the public sector and the imposing of *hijab* on female employees and clients of government departments and public services. The second stage comprised the imposition of *hijab* on a wider scale in public or private, whenever women were in the presence of men other than their kinsmen (Paidar 1995: 337). Despite its religious and legal aspects, the imposition of *hijab* was far from simple. While the revolutionary administration tightened its grip by using all means, including religious establishment, legislative, judiciary and executive powers to enforce *hijab*, a parallel informal group of committed revolutionaries took the law in their hands by using violence and harassment against women who did not observe *hijab*, and rendered impossible women appearing in public without it. The group justified its actions as the implementation of the Islamic tenet to "command right, forbid wrong" (*amr-i bi ma'ruf va nahy-i az munkar*), although the government discouraged it and religious leaders frowned on the violent physical abuse of women in public. Henceforth, the veil

> became a *de facto* national costume of Iranian women, when in 1983, the Parliament passed the Islamic Punishment Law (*qisas*) that stipulated 74 lashes for violation of the *hijab*. In response to women's continued opposition to re-veiling, in 1995, a note to Article 139 of the Islamic Criminal Code reaffirmed governmental penalty by mandating 10 to 60 days of imprisonment against those who publicly resisted the *hijab*. Thus concealing women's bodies, gender segregation and inequality became integral to state – building and its identity.
>
> (Sedghi 2007: 200–2)

Moreover, as Paidar (1995: 340) puts it: "the policy of Islamization of women's attire developed in parallel with public segregation of the sexes." The separation of the sexes in public was not an alien concept for Iranian society: it had been traditionally institutionalized and informally practiced for centuries. However, the Islamic state's enforcement of *hijab* and female–male segregation was lending these traditional practices a new political dimension. The two-fold process of sexual segregation of public spaces and, of social activities which were first implemented in public beaches, extended to other segments of the society. As noted in *Bamdad Daily*: "hairdressers were the next target, and those not acting quickly enough to segregate were threatened with confiscation of their income" (Bamdad 1979). This was followed by segregation in public transportation, schools, universities, etc. This practice met resistance from different groups: secular groups, who were against compulsory veiling; religious groups, who argued for the voluntary nature of religious belief and practices; and the government, who used various experimental and trial and error methods. Yet the government implemented the segregation of public activities such as political meetings and rallies, conferences, lectures and exhibitions. Images of women and men sitting in different rows, with a clearly observed distance in between, became commonplace. Parties, celebrations, wedding ceremonies and funerals, though private events, also had to have separate sections for women and men to avoid "sinful socialization." The fuzzy borders of public and private spheres under the political doctrine of *vilayat-i faqih* (the prevailing ideology of the Iranian post-revolutionary state, explained later in this chapter) provided the theocratic state means to extend its control over the private sphere. In short, as Paidar (1995) notes, the physical presence of women and men in the same space was strictly controlled and except for limited professional, educational and political reasons (such as highly specialized medical training with few attendants or women in male-dominated executive positions), and unless conducted under strictly supervised conditions, the mixing of women and men became a matter for criminal investigation and punishment.

The idea of building women-only parks as gender-specific urban spaces in Iran dates back to the mid-1990s, when it was suggested by the Deputy of Women and Family Affairs in the Iranian Presidency Office. Despite being in line with the Iranian government's grand gender-segregation policy, the idea was considered difficult to implement and remained neglected. However, the discussion about the women-only parks resumed once again in 2003 after the release of a medical research report commissioned by the Iranian Ministry of Health in which some women's health issues were partially and indirectly attributed to *hijab*. The report revealed the development of osteoporosis among a large number of Iranian women. The Islamic dress code for Iranian women in public spaces on one hand, together with small apartments with windows covered or blocked by thick curtains to observe the Islamic lifestyle and a sedentary lifestyle on the other hand, resulted in a lack of possibility for women to be exposed to the sunlight and a development of serious health issues for them. The research warned about the health hazards resulting from a lack of exposure to the sunlight and vitamin D deficiency among the coming generation of Iranian women. Islamic dress code,

in which only a small portion of skin is exposed, significantly reduces the surface area of the body exposed to the sunlight and hence diminishes the extent of
vitamin D intake. Among other known conditions, vitamin D deficiency is considered to be one of the main causes of Multiple Sclerosis, which has spread at an
alarming rate and has been pronounced an epidemic among the younger women
in Iran (Mahdavi Far 2015). Such concerns were the starting point for embarking
on the necessity of providing outdoor recreational and sport facilities for women.
Against these concerns, Tehran City Council in its 90th session on August 11, 2003
approved a bill and assigned the responsible bodies

> to respond to the needs of women in society – according to the Islamic and
> Iranian values, their physical, mental and social security – and to provide
> possibility of recreational, educational, cultural and social activities for them,
> enabling women to enjoy the benefits of direct exposure to sunlight at all
> age levels and ultimately, to enhance and improve the quality of individual
> and family life (through participation in collective and public spaces), five
> women-only parks must be located and constructed in Tehran.
>
> (Tehran Municipality 2003)

The ideological and normative-driven conditionality for presence of women in
public spaces in Iran restricted the options for meeting such needs. Hence, the
women-only parks were suggested as spaces to respond to the needs of women
for outdoor spaces and to lay a cornerstone for a free and yet conditioned public
space for women. At the next stage, the Council selected a committee to search
and suggest appropriate venues for realization of the urban women-only parks
in Tehran. Tehran Municipality as the executive authority assigned as the body
accountable for transforming and equipping parts of five urban parks in the city
to be used by women. Financial issues along with the lack of appropriate venues,
however (Fars News Agency 2004), forced the authorities to revisit the plan and
restrict their vision to build only three parks.

Two years after the first glimpse of the idea of women-only parks, in the 193rd
session of Tehran City Council in 2005, the Technical/Architectural Commission
of Tehran Municipality presented the proposition for potential locations of three
women-only parks in Tehran. The idea was approved and TPGSO received the
mandate to establish an auxiliary body to monitor financing, advancing, building,
supplying and maintenance of the entire women-only parks project. Meanwhile,
while working on the first part of the project, the municipality continued to
explore the possibilities for two additional women-only parks in Tehran (Fars
News Agency 2005).

In mid-2006, TPGSO revealed that the first two women-only parks in Tehran
(Pardis Banvan and Bustan Qa'im) would be inaugurated by the end of year and
that an additional two (Bihisht Madaran and Chitgar Park) were in their planning and construction stages, expected to be in use soon (Fars News Agency
2005). Despite those public statements, Bihisht Madaran as the first women-only
park in Tehran opened and started operating in 2008. Thereafter, three additional

women-only parks in Tehran were built and started operating, while parts of the Chitgar Park in western Tehran was also enclosed and equipped to be exclusively used by women.

Although the idea of women-only parks was not initiated for its ideological significance, it was implemented and flourished due to its consistency with Iranian post-revolutionary grand gender-segregation policy for public places. Even though it was embarked on because of an unexpected concern, i.e. medical complications for women, it did not take society by surprise. Many considered it the continuation of gender-segregation policy, which was implemented immediately after the Iranian Revolution of 1979, not least to mark the start of a new era. The gender-segregation policy, however, while restrictive in many senses opened up new possibilities for women of religious background. Meeting the requirement of the Islamic dress code and *hijab* in public spaces provided an opportunity for women to enter the public realm and claim professional and intellectual shares. The segregation policy was perceived by families with religious ties to provide a safe environment for women to participate in social activities. Many families, who formerly resented the presence of their daughters in institutions of higher education and gender-mixed working places, allowed and supported the female members of their families in their quest for education and a profession. Hence, paradoxically *hijab*, which was perceived restrictive by many scholars and activists, played a crucial role in facilitating the access of a given group of women to education and the labor market.

The systematic gender-segregation policy of the Islamic government started at institutions of education (from primary to tertiary, formal as well as non-formal) and extended to other areas of society. Dividing transportation facilities such as buses and subways – where the front door (or car) is used for men and the back door (or car) for women – post offices (Imrooz 2011), sport facilities (AKA Iran 2015), public pools (Vista News Hub 2015) and banks (Serat News 2009). The parks, hence, played a dual role of segregating women from public and providing them a public space of their own. Despite this, one shall not underrate the role of *'urf* (consuetude) in creating and maintaining the habit of visiting parks. While the rulings of *sharia* and Islamic law would not ban the presence of women in public spaces, and certainly less so if the space is women-only, the tacit rules of *'urf* may still overrule. A study commissioned by Tehran Municipality (Tehran Municipality Office of Social and Cultural Research 2011) shows that some 44 percent of women in Tehran would "never" or "a few times a year" visit the parks. Some 48 percent of this group perceive the parks as "inappropriate places," where "girls who fled home" and "drug addicts" are hanging out. Hence, many women, especially those who are particularly considered to be part of the target group for the parks, who are at the margin of suffering from a vitamin D deficiency and other symptoms due to a sedentary lifestyle, may abstain from using the parks to abide by the *'urf* and may regard going to the parks an unaccepted practice. Also, the spartan view on life, which is more prevalent among older generations of Iranian women, is likely to affect the contribution and presence of women in public spaces such as parks. This pattern of behavior, however, is subject to gradual change due to intergenerational interactions. There are instances of such interactions, where, for instance, a grandmother accompanies her granddaughter to a park (discussed extensively under social dimensions in Chapter 4).

On a closer look, *'urf* is a "Foucaultian notion of *episteme*. Somewhat like a world view, a slice of history common to all branches of knowledge, which imposes on each one the same norms and postulates, a general stage of reason, a certain structure of thought that a man of a particular period cannot escape – a great body of legislation written once and for all by some anonymous hand" (Arjmand 2008). It is in part realized through *tarbiya* (upbringing), the most important aim of the two-fold notion of education in Islam in the quest for *sa'ada* (prosperity). Through "the framework for the normative values," *'urf* also functions to maintain the status quo of power in society and it is reproduced by *savoir*, a knowledge exchanged through *talking about*. In *savoir*, the exchange of knowledge takes place through a process of socialization in various contexts. Hence, one can conclude that certain practices in public spaces – though not restricted by religion or law – are decided by unwritten rules of *'urf*, the dominant normative framework of the society. Meanwhile, the existence of a given tradition or practice in the society provides the person with *savoir* knowledge. Utilizing this argument in the context of this volume, one can come to the conclusion that the exchange of knowledge through the process of socialization within a given group and between various members of the group, along with the existence of such traditions in society at large contributes to the continuity and reproduction of given practices by women in public spaces.

Urban parks studied in this volume

The rapid industrialization of Iran, which reached its peak at 1970s, resulted in a massive rural to urban migration and drastic demographic changes. With an estimated annual urbanization rate of 2.07 percent (2010–2015), urban population in the country increased to some 73.4 percent in 2015 (Central Intelligence Agency 2015). This imposed a series of unprecedented complications in the country, more visible in larger cities. Rapid industrialization also turned air pollution into a serious problem across the country posing severe health hazards as well as social and urban problems. Rapid urbanization and invasion of the industries and modern constructions into the limited green spaces of an otherwise arid country such as Iran over decades has resulted in a scarcity of green spaces within the urban fabric. This lack of greenery around the multi-story rising complexes, along with the adoption of a Western lifestyle, has made the urban parks one of the most important components of a quality urban life in Iran.

The Iran–Iraq War, which started in 1980 when Iran was engulfed in chaos as a result of the revolution, lasted eight years and gained recognition as the twentieth century's longest conventional war. The war furthered the decline of the Iranian economy which had began with the revolution in 1979 (Karsh 2002: 19). As a result of the war, living standards dropped dramatically. The war swallowed almost all the revenue of the country and left the cities to expand without any plan or following any standard. At the end of this period (1980–1988) Iran was left with a heavy loss of infrastructure and human casualties, as is described by British journalists John Bulloch and Harvey Morris (Bulloch and Morris 1989: 239): "a dour and joyless place" ruled by a harsh regime that "seemed to have nothing to offer but endless war." After the war, and as part of the reconstruction plan (*dorey-i sazandigi*), Iranian cities started a makeover, not least to wash away the impact of war and destruction. The driving force behind many new "moderniza-tion" efforts was Gholamhossein Karbaschi a cleric with a degree in architecture and civil engineering, who assumed Tehran's mayorship from 1989 until 1998. He is widely acknowledged for his aggressive urban planning methods to improve the quality of life in Tehran. Among his efforts was the Jihad for Planting Trees (*jihad-i dirakhtkari*) to promote and benefit grassroots participation to expand the public green spaces. It was during Karbaschi's time that the urban parks received the recognition as being an inseparable part of the urban physical and social life. During his time in office, he inaugurated more than a hundred parks in Tehran and embarked on the construction of many more.

In the years to come and as a result of the steady economic growth and improve-ment of living conditions in Iran, the distribution of urban green spaces was considered to be one of the benchmarks for the quality life in Iran and attracted the attention of the urban planners. As the air pollution became a serious concern, Tehran Municipality took measures to partially meet the challenge with increasing the green spaces across the city to the extent that, between 2008 and 2011, some 424 parks of various scale were built in Tehran (Tehran Municipality's Statistics and Information Bureau 2015).

Whereas the efforts of Tehran Municipality in expanding urban green spaces and parks contributed significantly to the quality of urban life, the distribution of such spaces is far from even. There is a great disparity between various municipal districts in Tehran; and urban green space per capita varies from one district to next. While district 19 with 61.3 square meters per capita enjoys the largest urban green space (over twice the standard set for the city) some other distracts are struggling with a lack of green spaces. The data suggest that 6 out of 22 municipal districts (districts 7, 8, 10, 11, 12 and 17) in Tehran have less than 5 square meters per capita and fail to meet the international standard (Tehran Municipality's Statistics and Information Bureau 2015). District 10, with only 1.8 square meters, represents the lowest urban green space per capita in the city (see Figure 1.1).

In many ways Tehran – the economic and administrative heart of the country a metropolis with 8.2 million inhabitants (MAI, 2011), "is not an 'interesting' city. It is not like its regional counterparts, Istanbul or Cairo, with their long imperial or colonial histories, pivotal geo-political locations, memorable archi-tecture and natural charm. Tehran remains a provincial metropolis" (Bayat 2010: 99), "with streets choked by four million vehicles and air pollution that claims life of one person in every other hour" (Deutsche Welle 2015). "But it is a city with extraordinary politics, rooted in a distinctive tension between what looks like a deep-seated 'tradition' and a wild modernity" (Bayat 2010: 99). The idea of building women-only parks took shape in the midst of such chaos; the first

Figure 1.1 Urban green space per capita and location of the women-only and gender-mixed parks in Tehran.

Source: Tehran Municipality 2016.

women-only park was opened in Tehran in May 2008.[9] Many factors needed to be taken into consideration to meet the prerequisites for erecting such parks, not least the criteria of invisibility and concealment. A series of constraints including a lack of appropriate space for various activities, the high density of surrounding urban fabric, visibility of the site, safety issues and privacy concerns proved a challenge. Planners ultimately were forced to abandon the original idea of one park for each zone and were committed to locating parks in the most appropriate locations in the city, regardless of their proximity to a given zone. This shift in policy ultimately resulted in the establishment of five women-only parks in Tehran. The physical and morphological dimensions of the urban fabric in southern Tehran provided more possibilities than the northern part of the city. In this way, three out of five parks in Tehran came to be located in the southern part of city (see Table 1.1). Two out of five existing women-only parks in Tehran were selected for study in this volume, Bihisht Madaran in the northern part, and Pardis Banvan from the south (see Figure 1.2). Such division also follows the general socio-cultural pattern of the city. While the southern part is of a lower

Table 1.1 General information for women-only parks in Tehran

Women-only park	Opening date	Area (hectare)	Location in the city	Municipal zone
Bihisht Madaran	2008	19	North	3
Pardis Banvan	2008	27	South	15
Bustan Vilayat	2010	6	South	19
Bustan Qa'im	2010	17	South	18
Chitgar Park	2011	5	West	22

Source: TPGSO 2016.

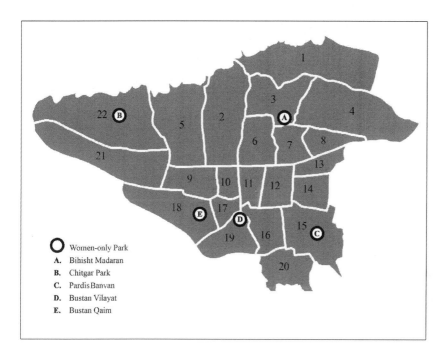

Figure 1.2 Location of women-only parks in Tehran.

Source: TPGSO 2016.

socio-economic background with more traditional and religious ties, the northern part is the home of socio-economically well-off inhabitants with more cosmopolitan tendencies and a lesser degree of religious attachment.

Women-only Bihisht Madaran Park in Tehran

Inaugurated in May 2008, Bihisht Madaran is regarded the first official women-only park in the country. With an area of 19 hectares (some 47 acres), the park offers a wide range of functions, amenities and services to its users. Bihisht Madaran was not originally designed or built for women; rather, it was transformed from a neighborhood park (previously called Nishat Park) into a women-only park. As one of the well-established parks in Tehran, Nishat Park was built in 1971 and served residents from its neighborhood prior to its transformation into a women-only park. In 2005, Tehran Municipality selected the park as an appropriate candidate for one of the women-only parks and renamed it Bihisht Madaran. Its natural characteristics and physical features, with some physical changes along the way, made this park one of the prime examples of a women-only park. The topographical composition of the park on the natural hills of Abbas Abad gave a solution to the visibility concerns. The geographical location of the park offered an additional advantage. It is located in one of the busiest junctions of the city, which provides high accessibility to the park and attracts women of various socio-cultural backgrounds to use it. Currently the park not only serves women but also uses women for the management and some maintenance.

Moreover, Bihisht Madaran with services and facilities such as a nursery, a café, shops, a library, an auditorium and a restaurant provides more than a green space for relaxation to its visitors. The park hosts females (and their accompanying boys up to five years old) for six days a week, from 07:30 in the morning to 19:30 in the evening. On Fridays and official holidays, some parts of the park are open to the gender-mixed general public and are used mostly by families and groups as a popular picnic destination.

Women-only Pardis Banvan Park in Tehran

Pardis Banvan was inaugurated in August 2008 as the second women-only park in Tehran. With an area of 27 hectares (some 67 acres), this park is the largest and the first specifically designed women-only park in the country. With its large area and diverse range of available activities, Pardis Banvan aims to serve users beyond its neighboring districts. The park provides various activities and services, and once inside offers a safe, secure and relaxing milieu. Compared to Bihisht Madaran, Pardis Banvan houses a wide range of amenities and activities both open air and indoor. Water facilities including swimming pools, sauna, Jacuzzi and sunbathing are among the most popular activities, and attract many women.

Pardis Banvan is located in an island-like enormous lot, previously vacant and not much suitable for many other urban functions. It was seized by the municipality to design and establish a women-only park. The location met the main criteria of invisibility set for women-only parks by policymakers and planners. It is located in a low-density urban fabric and surrounded by various functions, although not all of them compatible, mostly of urban functions of some kind.

The park is intended to "provide women access to a safe space for various activities … and direct exposure to sunlight, which has an important role in curbing diseases including osteoporosis" (TPGSO 2015). Hence Pardis Banvan with its numerous activities is not only a green space but also a complex for various activities. According to the promotion materials by the municipality, activities are planned in four main categories:

- educational activities, including courses and workshops on pottery, carpet weaving, cooking, table setting, computer, religious training, and foreign languages;
- athletic activities, including bodybuilding, swimming, sauna, Jacuzzi, basketball, biking, horse riding and motorcycle riding;
- services: the lake, flower conservatory, toddlers' playground, auditorium, shops, restaurant and parking; and
- other facilities, including spaces for mothers with infants and seasonal food markets.

The park is open daily from 08:00 to 19:30 for females and boys younger than five years old at the company of a female visitor and is closed on Saturdays.

Gender-mixed Niavaran Park in Tehran

Located in northern Tehran, at the neighborhood of the same name, Niavaran Park is one of the oldest parks in northern Tehran. Niavaran neighborhood – an area with residences of higher cultural and socio-economic status – is in fact a mixed district (in terms of variety of land uses), which mostly provides residential and commercial complexes. Geographical proximity to the Alburz Mountains bestows a beautiful environment and fresh air for an otherwise polluted Tehran. In addition to its natural and economic capacities, the park is located next to *Farhangsara va Kakh muzeh-yi Niavaran* (Niavaran Cultural Center and Palace Museum), a historical and cultural complex. Niavaran Park is a popular public space in the district, which offers a wide range of facilities and services.

With an area of 6 hectares (14 acres), the park was designed by British architects in 1963 and opened to the public in 1969. Located on a hillside, Niavaran Park has made creative use of the difference in height levels to provide a more efficient space with a variety of views. Various spaces within the park are connected through stairways to provide a beautiful and unique landscape. With two main entrances at the northern and southern corners, the park is surrounded by old tall trees and bushes which are also used as a soft barrier to mark the borders of the park and separate the park from neighboring spaces.

Gender-mixed Bi'sat Park in Tehran

Bi'sat Park is one of the oldest urban parks and officially the first recreational and sport complex in southern Tehran, an area with distinct social and cultural characteristics. Popular among local residents, Bi'sat Park has not only affected the social life of the surrounding community, but also the air quality and urban landscape of the area. The park is a large open urban space with a naturalistic designed landscape which is separated from surrounding urban fabric by short walls. With the area of 53 hectares (some 131 acres), the park provides various types of facilities and activities. It is located in a dense and crowded district of the city and attracts various groups of users into the park. Several entrances in different directions of the park give the neighboring residents and other users easy access to the park and its amenities.

Women-only Pardis Park in Isfahan

Known as an open-air museum of Islamic and Iranian art and architecture, and with a population of some two million (Iranian Bureau of Statistics 2011), Isfahan is the third largest city in Iran. Hitherto (August 2015) there are five women-only parks actively operating in the city: Bagh Nush (Municipal Zone 12, 2 hectares), Pardis (Municipal Zone 10, 1.8 hectares), Nazhvan (Municipal Zone 9, 2.8 hectares), Sadaf (Municipal Zone 13, 6 hectares) and Tulu' (Municipal Zone 15, 2 hectares) (see Figure 1.3). In many ways, the popularity of the women-only parks in Isfahan is an indication of the stronger religious and traditional commitments among the inhabitants of Isfahan. Isfahan also seems to be the leading municipality in the country on building and promoting women-only parks. This, in turn,

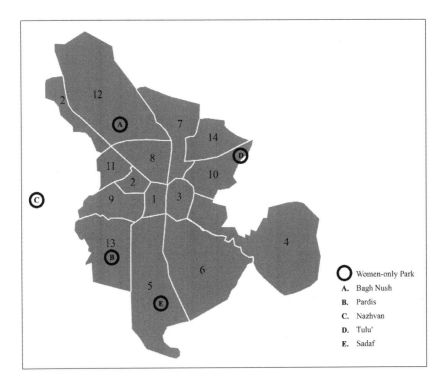

Figure 1.3 Location of women-only parks in Isfahan.

Source: Isfahan Municipality's Digital Portal 2015.

is a part of the municipality's policy to expand the green spaces in the city as a marketing strategy to present Isfahan not only as one of the main tourist resorts in Iran for its historical sites but also for its various recreational alternatives.

Inaugurated in 2002, Pardis Park in Isfahan with an area of 1.8 hectares (some 4.5 acres) seems popular and well recognized among the Isfahani women. The park is located in the western edge of the city, near the Nazhvan woods, a green belt surrounding the city and the bank of the Zayandehrud River, which crosses through the city. The venue of the park was originally a private garden, which was purchased by Isfahan Municipality and transformed into a public women-only park as part of the overall policy to build gender-separated public spaces.

Compared to those in Tehran, women-only parks in Isfahan are significantly smaller with limited resources. However, they are engaged in similar types of issues and concerns as larger women-only parks in Tehran. In order to speed up the process of building women-only parks, Isfahan Municipality found the solution of creating smaller parks and dealing with the administrative and technical problems through decentralization and local-level decision-making. Constraints – including the criteria of invisibility – also forced planners and authorities to move the parks to the outskirts of the city. This, however, has affected the access of women to those parks and meant that the parks are not evenly distributed across the city. Four women-only parks located in the southern part of the city, within a close distance of each other, while there are no women-only parks in other parts. Pardis Park in Isfahan is surrounded by farmlands and residential functions. Compared to two women-only parks studied in Tehran; they offer a limited range of amenities, services and activities.

The amenities in Pardis Park include volleyball, basketball and badminton courts, a skating rink, a biking track, bodybuilding facilities, a shop, an exhibition hall and an auditorium for 80 people. The park operates from 08:30 to 17:00 everyday.

Women-only Park Banvan in Rasht

The city of Rasht, with 700,000 inhabitants, is regarded as a medium-size city by Iranian standards and is one of the most popular tourist resorts in northern Iran on the coast of the Caspian Sea. With its humid subtropical climate characterized by hot, humid summers and generally mild to cool winters, Rasht is a hospitable climate for vegetation. Despite this favorable climatic condition and natural characteristics of the city, the urban green space per capita for Rasht is considerably lower than the standard level defined for such climate. One may argue that the rich green nature available for inhabitants may satisfy such needs, especially for purposes such as outing and picnics. Currently, the city has 40 parks, one of which is women-only and located in the city center, expected to serve women from all around the city.

With an area of 2 hectares, Park Banvan in Rasht is the only women-only park in the province of Gilan – with an area of 14,042 square kilometers and population of some 25 million (in 2011). Park Banvan is different from all three parks described earlier, in terms of both form and function. It also introduces a new category of women-only park in Iran. At first glance, the park resembles any gender-mixed park in the city, surrounded, however, by a net fence. This means that the mandatory veiling (*hijab*) must be observed at all times when one is inside the park. A plaque at the main entrance is the only sign to reserve the park for women. Park Banvan is located in downtown Rasht at the heart of the business quarter of the city, within a rich urban fabric and located alongside the river which crosses Rasht. The park's vicinity to the university, neighboring schools and sport facilities contributes to its visibility and increases the number of visitors. With its symbolic entrance gate, Park Banvan is open and available around the clock.

Notes

1 English interpretation by Shahriar Shahriari (see Hafiz 1999).
2 *Bagh*, the Middle and New Persian word for "garden," and also the Sogdian *βay*, strictly meant "piece" or "patch of land," corresponds to the Gathic Avestan neuter noun *baga*- "share," "lot" (Bartholomae 1904). The word with its derivatives is also used as an alternative term for parks in Iran. The old word for "garden" *paridaiza* (Old Persian *paridayda*), literally "walled" (whence *pardiz*, Greek *ho parádeisos* "park for animals," "paradise," Arabic *ferdaws*) survives in the New Persian *paliz* "vegetable garden," "melon bed," though today this most often denotes an unenclosed patch. Other words for "garden" are the New Persian *bustan* (from *buyistan*, whence the Armenian *burastan*), literally "place of perfume," Arabic *bostan* (plur. *basatin*), and *golestan* "rose garden" or "flower garden" (Eilers 1974).
3 Modern interpretations of such a design are utilized in Niavaran Park at Northern Tehran and Pardis Banvan, two of the parks studied throughout this volume. Niavaran Park is designed on a classic Persian garden, using uneven surface and natural slope of the area for the distribution of water and vegetation as well as the amenities and activities across the park.
4 This phenomenon is rather common in other parts of the Muslim Middle East. For other instances see al-Harithy 1994.
5 '*Urf* refers to the custom, or "knowledge," of a given society. When applied, it can lead to the deprecation or inoperability of a certain aspect of *fiqh* (Islamic jurisprudence) (Glenn 2007: 201). '*Urf* is a source of rulings where there are not explicit primary texts of the Qur'an and *Sunnah* specifying the ruling. It can also specify something generally established in the primary texts. Essentially, Islamic law resolved the tension between theory and practice by *de facto* recognition of the role of custom. This was achieved by several devices. One was resorting to the other legitimate sources of law. A particularly important principle in this context is personal preference of a jurist (*istihsan*) through which a given customary practice could be incorporated into the law (Libson 2012; Libson 1997). Hence, contradiction between '*urf* and *sharia* arises in certain occasions. There are instances where a given practice is not against *sharia*, however, it is conditioned or banned by the (sometimes even) tacit ruling of '*urf*.
6 *Namus* is a concept of an ethical category, a virtue of patriarchal nature, widely used in a strong gender-specific context of relations within a family described in terms of honor, respect/respectability and modesty. It is important to note that the concept of *namus* in respect to sexual integrity of family members is an ancient, exclusively cultural concept which predates Judaism, Christianity and Islam. It is claimed that religious alignment with *namus* does not exist in any of the holy scriptures of these religions.
7 Although in some parts of the country oriel covered windows – most likely inspired by Arabic style *mashrabiya* – was used to cover the windows of the *andaruni*.

8 The author has borrowed "generification" from linguistics where it is employed as the process of using specific names or brand names of products as names to indicate a genre or specific group of products in general and takes it into another level of abstraction at theoretical level.

9 An administrative officer in Pardis Park in Isfahan in an interview with the researchers of this volume claimed that Pardis Park in Isfahan was the first women-only park in Iran (established in 2002). This is also recorded on an information plaque at the very entrance of the park. However, the government's countrywide women-only park project was formally inaugurated in 2008, with Bihisht Madaran Park in Tehran being the first women-only park in Iran.

2 How physical and morphological dimensions affect the female body in the public space

Women-only versus gender-mixed parks in Iran

with Masoumeh Mirsafa

> The city's form and structure provide the context in which social rules and expectations are internalized or habituated in order to ensure social conformity, or position social marginality at a safe or insulated and bounded distance (ghettoization). This means that the city must be seen as the most immediately concrete locus for the production and circulation of power.
>
> *(Grosz 1992: 250)*

A cross-disciplinary study of gender and space informs a methodological approach beyond the traditional disciplinary boundaries. Recent research inspired by feminist scholars across various disciplines has influenced the cross-disciplinary research to shift the attention from the notion of space as it has been defined by urban planners and architects to the notions of agency, representation, power dynamics and spatial metaphors as it is defined, used and transformed by actors through everyday activities and practices (Rendell *et al.* 2000; Rieker and Ali 2008; Spain 1992). Such cross-disciplinary approach in this volume is partially based on the active role of urban planning and architecture in the construction of sexual identity, and on how the architects, artists and theorists investigate the way sexuality is constituted through the organization of materials, objects and human subjects in the actual space. Positioned within such a discourse, Sanders (1996) informs the question of sexuality and space under an irrevocable account on the history, context, theory and practice. This also is in line with what Spivak (2003: 26) calls "looking for the definition in the eye of the other," calling for a shift from urban planners as the authority of knowledge to those who actually use the space.

The methodological approaches in studying space as informed by urban planning suggest looking into space using a multi-fold scheme composed of various dimensions. The classification perpetuates a divide to address mostly the designers' intentions *vis-à-vis* the actual usage of the space to include morphological, perceptual, social, visual, functional and temporal dimensions of a space. Whereas such multi-level analyses correspond to an in-depth understanding of the space in urban studies/planning, it proves too detailed an account for an interdisciplinary study to permit to communicate effectively between disciplinary boundaries.

Against this premise, under the rubric of morphological and physical dimensions, this volume endeavors to address morphological, perceptual, visual and temporal aspects of the women-only parks and gender-mixed parks in Iran. Given the fact that women-only parks are closed spaces in which carrying any device to record any visual footage is strictly prohibited and is subject to a law suit, an ethnographical thick description approach (Geertz 1973) is used to provide an all-encompassing and detailed portrait of those parks. Such an approach is composed of what Geertz (*ibid.*) labels a microscopic description of the specific and contextualized happenings that serve as an ethnographical miniature. The notion of thick description is composed of the detailed explanation and analyses of the context, its setup, and the human behavior which is meaningful in such a context. Hence, the contextual analyses are described under morphological and physical dimensions in this chapter and functional dimensions in Chapter 3, whereas the social and behavioral dimensions will be discussed in Chapters 5 and 6.

Built as an analytical tool suggested by scholars of urban studies such as Rapoport (1969), Carmona (2003), Hillier (1996) and Lefebvre (1991), the morphological and physical dimensions describe, study and analyze the layout and configuration of urban form and space. It includes those spatial and aesthetic features, which promote the quality of social life, and reflects the intended functions of the place on the one hand and the actual usage of it, on the other hand. The suggested analytical tool is modified to meet the context-specific characteristics of the parks. Hence, the morphological and physical dimensions in this volume include an in-depth thick description of accessibility (city-wide and in-park), legibility (both in macro and micro scales), enclosure and permeability, and visual attractiveness. After a description based on extensive observation sessions, analyses of the policy documents, and review of reports and interviews with the authorities and users both at the women-only parks and with women in gender-mixed parks, a detailed account is provided to compare women and their interactions (social and otherwise) in both urban spaces.

Accessibility

Creating accessibility has always proven a challenge in designing public urban spaces. An urban space is regarded accessible when the target population is able to access its intended amenities and resources without any major obstacle. Accessibility also means providing the possibility for participation in day-to-day activities involving community life. Lynch (1960: 118) defines access as the ability to reach a diverse range of persons, activities, resources, services, information or places. According to DETR (2000), access is "the ability of people to move around an area and to reach places and facilities." The possibility of freely moving around and using the resources and services within a given space, as well as the ability to reach the place without any major complications, is central to both definitions.

Furthermore, the possibility for activities in pleasant and pleasing environments, such as those found in urban parks, popularizes these places for people who live in such a large, crowded and polluted metropolis as Tehran. Going to nearby parks, doing outdoor activities, and enjoying the fresh air of a wider green space would be a much better alternative to staying at home in usually small apartments and condominiums. In addition, it may provide possibilities for those who prefer to escape the crowd, pollution and other complications of urban life momentarily. While public parks make urban areas more inviting for living, working and relaxing, the most basic problem for people in using public open spaces, such as parks, is the accessibility of them. Whereas for certain groups of users (e.g. mothers with baby carriages or users needing wheelchairs) limitation of access can be a major hindrance; a park's location and its ease of access is a main decisive factor for all users.

In such public spaces as parks, the park's location is considered a pivotal factor in making a place accessible. Location is often regarded in terms of linkage, connectedness, walkability, vicinity and transportation. Access through the convenience of public transportation is a major factor in designing parks and other public spaces. Urban spaces that are only accessible through private means of transportation like cars are considered inaccessible in urban design. Access through public transportation becomes even more important when public spaces are intended for potentially vulnerable, socially marginalized or economically disadvantaged groups within a society, such as those with limited access to private means of transportation. Generally speaking, women-only parks as public urban spaces are expected to be easily accessible to its intended group of users – namely women.

Currently, of five women-only parks in Tehran, three of them are located in the southern half of the city. Only the Bihisht Madaran Park is left to serve all women from the northern half of the metropolis with some eight million inhabitants. Bihisht Madaran is located in the city's functional core, thus providing good access from different directions and attracting many users from distant corners of the greater Tehran area. While many users do commute for a number of hours to use the park, the long distance to access Bihisht Madaran and its facilities is regarded as a major problem. As a visitor of Bihisht Madaran puts it, "There is a need for more parks in Tehran. It's impossible to come here frequently as it is really far from my place." Accessibility of public transportation is regarded as one of the main factors affecting the selection of any park among the Tehrani population, to the extent that some 82 percent of respondents in a recent study consider that it is a "very important or important" criterion in their decision to use a park (Tehran Municipality Office of Social and Cultural Research 2011).

Observations of the four women-only parks studied in this volume reveal a continuum of variations in accessibility to their locations and services. Bihisht Madaran is located in the central part of Tehran, and is surrounded by natural hills from the west and residential areas from east. It has good connectivity to the surrounding urban fabric and to the transportation network in Tehran. The continuity of the residential fabric and its connectedness to the park provides a short walking distance for women who live in the neighborhood, something that is perceived as a great advantage for them. Except for its western and southern sides, the park has simple and rapid access to public transportation (bus and subway). There are also facilities for those who prefer to use private cars.

Tehran's Pardis Banvan women-only park's location and accessibility is completely different to that of Bihisht Madaran's. Visiting the park means facing difficult access and use for those living outside the neighborhood area. Long commutes from and to any transportation facility, a lack of appropriate route to the city, and heavy traffic load make it difficult to reach this park. Furthermore, with the lack of a subway line close to the park, visitors have no option but to use their private cars or to make multiple changes using buses and taxis in order to access the entrance.

Walking around Pardis Banvan's outer walls provides a better impression about the urban fabric of the area where the park is located. There is no possibility of accessing the park on foot. The aerial view of the park shows how isolated it is from the urban fabric with highways passing by in different directions. These highways act as separators, which shape a border around the park and neighboring public space (including a water park in the south and a bus terminal in east) to make the area even more isolated. Several women noted a lack of public transportation as a serious limitation for Pardis Banvan. One woman said,

> We always come to the park with our private car. It would be very nice if they could have more buses to transport people to different parts of the city. There is a bus stop outside the park close to the main entrance but it just goes in one direction which is the opposite direction to our home.

From an urban design perspective, for pedestrians the connectedness between various public and social spaces and public space being integrated within a surrounding fabric is of pivotal significance. Most of the women-only parks, however, are located in suburban areas to maintain the privacy and invisibility of the place, as is required by the planners to meet the religious/ideological criteria of concealment. This has resulted in low accessibility for pedestrians. This is especially the case for Pardis Banvan where the highways and city-scale functions have created a less pedestrian-friendly environment.

Pardis Park in Isfahan is part of that city's western suburb. It is surrounded by a main highway to the east and the naturally grown woods of Nazhvan to the north and west. As the highway is located between the park and a residential neighborhood, it functions as a physical barrier. Even though it is only a walking distance to the park, there is no way to access it other than by crossing two bridges over the highway close to the park's entrance. Under the existing situation, and with a lack of access to public transportation, private car or taxis seem to be the only alternatives to access and to use Pardis Park. The difficulty in access is a result of a requirement for where women-only parks in Isfahan may be located. This requirement states that these parks should not be visually accessible from neighboring buildings. Thus, due to the intense urban composition and lack of possibility of placing parks in the inner parts of the city, all women-only parks in Isfahan are located on the city's outskirts and close to highways or in yet-to-be-developed areas. This is also one of the recurrent themes brought up by park visitors. Hence, access to the park remains a main concern for park users to the extent that, if they had any other alternatives or possibilities, they would prefer not to come to these parks.

Park Banvan in Rasht provides a good example of an accessible urban space which is located in a central part of the city, which has great access to the surrounding areas, and which has good connectivity with the city's transportation network. The park is located in a residential area of the city and is very close to a university campus, making it a popular venue for female students during their breaks.

Concerning the criteria of accessibility, DETR (2000) suggests that it is necessary for an urban space to be available to a wide range of users including elderly, disabled, children and other groups of people in need of assistance. Sircus (2001: 31) maintains that in designing public spaces special attention should be given to "women and low-income groups who have reduced mobility and access because they rely on public transportation." If a place lacks sufficient functioning public transportation facilities, despite being attractive, it would be difficult for women to use it. In the words of a park user in Bihisht Madaran, "once I used to attend outdoor aerobics classes and I liked it a lot, but I couldn't continue that. It is simply too far to where I live and it's too complicated to get here."

This, in turn, proves Arefi's (2014) point that an integrated and well-functioning transportation system encourages people to use public transportation to move between various places. It can also give people incentives to come out of their homes and take part in public life. All types of movements from cars to walking, biking and public transportation should aim to meet the needs of the widest range of people possible. Furthermore, provisions for functioning public transportation must be met along with facilities for private cars. To encourage private car users to choose public transportation, the criteria of access, efficiency and convenience must be met, since as Sheller and Urry (2000: 749) put it, "cars afford many women a sense of personal freedom and a relatively secure form of travel in which families and objects can be safely transported, and fragmented time-schedules successfully intermeshed." For those who can afford them, cars bring security and peace of mind and unless a convenient alternative is provided they hardly ready to use other means.

The accessibility of a place is also in part affected by its "permeability." A place is regarded accessible if it is permeable. Permeability is "the degree to which an area has a variety of pleasant, convenient and safe routes through it" (Cooper *et al.* 2009: 146). For Bentley *et al.* (1985: 12) the quality of permeability is, "the extent to which an environment allows people a choice of access through it, from place to place," which includes both "physical" as well as "visual" access. The routes are key in creating physical permeability, and connected routes increase the access alternatives and give users more possibilities within a given space, while visual permeability allows people to see into a place before entering it in order to judge whether they would feel comfortable, welcome and safe there.

The requirements for constant surveillance and strict control pose a series of limitations on the permeability of the women-only parks. Physical permeability in women-only parks has to be provided at entrances. However, the parks are enclosed areas mostly with a single entrance point, which gives yet another restriction to their permeability.

Observing different women-only parks in Tehran and two other cities shows variations in the permeability of the parks. Bihisht Madaran, Pardis Banvan in Tehran, and Pardis Park in Isfahan are gated spaces and lack the possibility for visual access to the inside of the park. Physical connection with the outside is only possible through the parks' main entrance(s). Park Banvan in Rasht, however, is entirely different. As noted earlier, the park is not enclosed (gated) and has high permeability, both physical and visual.

The large size of the parks and lack of appropriate or limited points of entry is one of the constraints mentioned repeatedly by the park users. Pardis Banvan, with an area of some 27 hectares, has only one entrance. Bihisht Madaran, on the other hand, with its 19 hectares has three entrances. The lack of entrances to these parks, which have such a large surface area, has resulted in an uneven distribution of the facilities and services. Many users prefer to stay close to the entrance(s) rather than moving around and distancing themselves. Whereas some parts of parks are overused, other parts are hardly visited or used at all. Time and again, park users point this out as one of the main obstacles to enjoying the parks' full capacities. A female security officer in Pardis Banvan brings this up from a different perspective:

> This park has just one entrance, which is unacceptable for such a large park like this one. I live close to the park, if the park had another entrance on the northern or eastern side, I could have easily entered the park as I got off the taxi or bus. It would have saved me a lot of time. But now I have to walk this ring path around the park to access to the main entrance.

Strict regulations backed by religious normative values, however, rank the privacy of the parks as the highest priority. As women-only public spaces, the parks have to be controlled and remain under constant surveillance. Limiting points of access, although against urban design principles, provides better means to tighten control of the parks and, therefore, it has been implemented. This control also contributes to the lack of permeability and affects the internal connectivity and access to facilities and services, which in turn influences the vitality.

Also, the needs of women with limited mobility are not considered in these parks. Iranian disability law requires all public places to accommodate the needs of individuals with mobility limitations, and to provide them with access to use the facilities and services. Despite the law, many public facilities lack such services mostly because at the time of their construction the notion of disability was not the focus. Parks are among the public places where creating access is easier than in many other spaces because they are usually built at ground level, which makes it easier to enter the space, and since they are usually permeable from multiple points. Despite this, according to a member of the Tehran Council for Improving Public Spaces, "a new park is built in Tehran each month, but no provisions are made to ensure the facilities are accessible to persons with disabilities"

(Radio Zamaneh 2013). While access for people with needs or limitations is of significance for parks in general, it is rendered even more important for women-only parks since mothers with baby carriages are among the main target groups of users for those parks.

The topographic characteristics of a park are also an important factor in providing access for those with physical limitations. In general terms, parks located in the plain areas (Pardis Banvan and Bi'sat Park in Tehran, Pardis Park in Isfahan and Park Banvan in Rasht) usually accommodate those needs better than those at the hill slope (Bihisht Madaran and Niavaran Park).

Accessibility of the women-only Bihisht Madaran Park in Tehran

Bihisht Madaran was designed as a local park and has been functioning (though not specifically targeting women) since 1987. Known previously as Nishat Park, it is situated in a central part of the city and in the middle of a commercial and residential area. It has been successfully integrated within a dense urban fabric and its surrounding streets help to facilitate access to the park. Haqani Freeway connects the northern part of the city to the park and makes the park available to women from northern Tehran. In the eastern part, a number of smaller streets connect surrounding residential areas to the park. Almost all the neighboring streets lead to the park and are pedestrian friendly. This makes the park easy to access for inhabitants living in the neighborhood. Furthermore, Bihisht Madaran enjoys city-wide fame with women traveling from all over the metropolis to use it. As it was planned for the park to be used by women from all over Tehran, accessibility becomes an important factor. The park has very good accessibility for pedestrians and through all means of transportation including private cars, taxis, buses and the subway.

Since the park is situated at the heart of a residential and commercial area, city planners have been able to provide good connectivity and accessibility to it from all over Tehran, as well as to provide visitors with different alternatives for reaching the park. Also, the part of the city where Bihisht Madaran is located has a rather old and well-established urban fabric where all means of transportation have been built up over time. The presence of such crowd-intensive institutions and activities in the neighborhood, including Musalla (a prayer congregation center) and the Arasbaran Cultural Center, among other institutions, have contributed to this development and intensified the need for accessibility and connectedness. When the park was originally established, designers used the area's natural topographical features creatively. As the park is located on a hill, which limits access from the northern side, designers endeavored to provide an entry point from the north but kept the slope as a natural hindrance to close visual connectivity as they turned it into a women-only park. This led to the construction of a bridge over the freeway, intended for use by those trying to reach the park from the nearby metro (Himmat) and bus stations. The western side of the park was originally meant to be accessible from multiple points. However, later, when the park was changed to a women-only park, they closed all entrances and kept only one.

In total, the park has three entrances. The eastern entrance is meant to, and usually does, attract local residents who mostly walk to the park. There are no public transportation facilities close to the eastern side. Although mostly overcrowded, there is a possibility of finding a random parking spot for those who drive cars. As many women argued in their interviews, an easy and convenient access to a parking lot close to the park increases the possibility for women to drive to the park and reduces their dependency on the male members of their families to give them lift; there is not any designated parking spaces or any public parking lot available here. Although women can park on the sides of the streets, the proximity to other key urban functions such as companies, businesses and the Arasbaran Cultural Center have created a concentration of local visitors in this area. The single entry point, which was introduced when the park was transformed to women-only, has made the access more complicated. Whereas, previously, one could park the car at any given spot and enter the park, now one must sometimes walk a long distance to access the entrance. The designated entrance opens to the middle of the park, which in turn provides better accessibility. Also, because this entrance connects the park to the surrounding residential fabric, there is always the possibility of finding services (such as shops and cafés) outside the park to replace those inside, should visitors wish to use alternative facilities during their time there.

The second entrance to Bihisht Madaran, which is the northwestern entrance from the Haqani Highway, is more convenient for those who arrive by their own private cars. The park is more accessible for cars to use because there is a free parking lot at the entrance. Since the way to the park is on a steep hill, the parking lot close to the entrance proves useful and contributes to accessibility. This entrance is also used by those who arrive by taxi. While rather convenient for cars, access to public transportation facilities is limited at this northern entrance. The closest bus line is some 500 meters away, and the metro station, which is connected through a crossover bridge, is neither easy to use nor close. Therefore, while the metro has proximity to Bihisht Madaran, it is not convenient enough for visitors to use to reach the park. It seems that, in designing this entrance, the target groups focused on primarily were private car owners or those who prefer private means of transportation such as taxis.

Furthermore, this entrance is not close to facilities and services found inside the park. Hence, the entrance is mostly used by those who use the park for picnics. It is easier to carry things inside the park, since a car can stop very close to the entrance and since the parking at this entrance is not usually crowded. Due to its distance from facilities which attract users and which create noise and crowding, this part of the park is calm and more appropriate for picnicking.

And, finally, the southern entrance from the Himmat Freeway provides the best access to the park from public transportation. There is neither access to parking nor the possibility to park a private car at this entrance. However, there are shuttles which bring visitors from different parts of the city and drop them off at this entry point. Arriving by car at this destination, the driver has to unload passengers and then drive further to park at the northern parking lot. There is also a bus stop at the very entrance and, while not very convenient, there is access to the park here with the metro. Those arriving by metro prefer it over the traffic jams that they would otherwise experience using cars or buses.

Although Bihisht Madaran is located on a hilltop with a rather sharp slope, access for users with disabilities and mothers with baby carriages, even if restricted, is not impossible (at the main entrance). Access is available for this group because the slope is not inclined with stairs, and its gradual pitch allowed designers to provide an access ramp for people with limited mobility or special needs. At some areas, the slope's angle is sharper than 12 percent, the maximum recommended threshold to allow mobility for people with disabilities. This has caused access difficulty for people in need of aid equipment. No specific equipment or services (such as elevators or vehicles) are provided at the park for people with limitations or specific needs.

To sum up the notion of accessibility at Bihisht Madaran, one can generally conclude that the northern entrance is meant to be used by visitors who prefer private cars; the eastern entrance favors pedestrian use; and the southern entrance is for those who use public transportation, including bus and subway (see Figure 2.1).

Accessibility of the women-only Pardis Banvan Park in Tehran

Whereas Bihisht Madaran could be regarded as an example of a park with good accessibility, Pardis Banvan in southern Tehran is an instance of a park with poor access. In reality, Pardis Banvan resembles an island surrounded by a network of freeways and highways, and is detached from the urban fabric. The park's immense size, along with its oval shape, has also contributed to this isolation. The bare environment around the park, which makes it impossible to find shade on a hot summer day or refuge on a rainy autumn one, makes it literally impossible to reach the park on foot except from on one single side. As the park is situated in a very different urban fabric than that of Bihisht Madaran, walking to the park is an unpleasant experience. The park is only available from its southern side, which is a more residential area. The proximity to two other local gender-mixed parks at this side does make a certain type of access somewhat easier. There have been instances, for example, of families coming to the park for a day outing, but at the entrance they separate. The women continue on their path to Pardis Banvan, while male family members head to Azadigan Water Park, which is designated solely for men. Moreover, the lack of complementary urban fabric in the neighborhood surrounding the park creates a sense of uneasiness and feelings of being unsafe for many women.

Since Pardis Banvan was originally designed as a women-only park, isolation and possibility for enclosure were among the primary criteria for selecting its venue. The location meets those requirements because it is far from any amenities or functions, making it easier to enclose and control the park, and because it stands separate, making isolation possible. This means that there was no need for the park's designers to take any specific measures for concealing the sides of the park, since there is no function to overlook the park from those sides. The problem

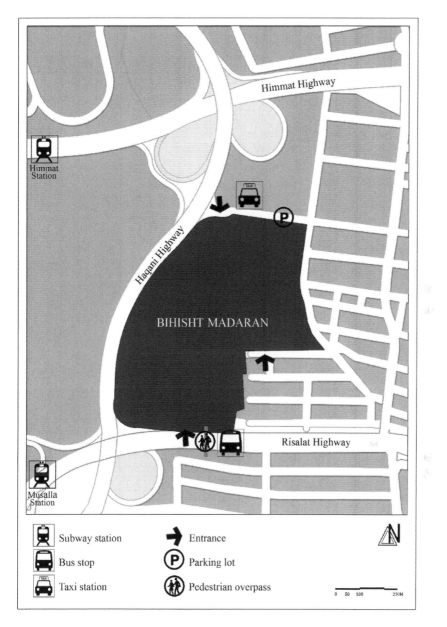

Figure 2.1 Accessibility by various means of transportation to the women-only Bihisht Madaran Park in Tehran.

with the accessibility at Pardis Banvan becomes even more evident with the realization that it has only one entrance even though it is the largest women-only park in the city with an area of 27 hectares.

While Pardis Banvan is accessible through all means of public and private transportation, it is not free from hassle. Shuttles and taxis mostly drop visitors off at the park's northern corner where a large taxi terminal is situated. This terminal is at a rather long walking distance from the entrance, which becomes an unpleasant experience under extreme weather conditions or under feelings of being unsafe in an isolated area. There are also shuttle taxis that stop at the entrance; however, they are not frequently available and visitors are sometimes forced to wait for quite some time to get a shuttle (see Figure 2.2). Female visitors are highly advised, through various signs and also by park guards, not to trust vehicles other than authorized shuttles (distinguishable by their color and signage). This, in turn, adds to the unsafe feelings many women experience, not because of exposure to harassment or violence but because of the many unsubstantiated stories women have heard and which have resulted in the park having a bad reputation. Thus, women rely on shuttles which take passengers to the closest metro station.

Figure 2.2 Accessibility by various means of transportation to the women-only Pardis Banvan Park in Tehran.

Pardis Banvan is also accessible by private cars. There is a free parking area close to the park entrance which was built to serve visitors arriving by car. As the area is not usually crowded, there is always available parking space close to the entrance. There is also a bus stop right at the park's entrance which connects it to the Khavaran Bus Terminal, a transportation hub in southern Tehran with services to many areas in the city. Along with private cars, the bus is a convenient means of transportation for reaching the park. There is also a metro station in the neighborhood (about 10 to 15 minutes walking distance); however, because of feeling unsafe, visitors coming from parts other than the neighboring area do not use it.

As noted earlier, *feeling* unsafe was repeatedly brought up by many women and results in preventing them from using the subway or walking through the neighborhood to reach Pardis Banvan Park. Many women have never experienced any type of harassment; however, they do mention the presence of men on the street as being an unpleasant experience. "Walking through a sidewalk in which men are sitting on both sides" is an encounter that many of these women would prefer to avoid. There have been cases reported of verbal harassment, a problem which is visible across the city in general, but it is hard to determine whether it takes place here more often than other parts of Tehran.

While conducting observations on the park, it became noticeable that many women are brought to the area on motorbikes driven by their husbands. As the park is situated in the southern part of Tehran, an area with inhabitants of lower socio-economic status, many are without cars. Therefore, these husbands drop their wives off at the park entrance and continue to drive to work or run errands. They later pick their wives up on the way home, or the women return on their own using public transportation. That said, women visitors traveling by family owned cars are usually driven by men taking them to and from the park. Additionally, there are young women who drive their mothers or older female family members to Pardis Banvan and then return to pick their deposited members up after a specified time. Women's dependency on others to provide them with access to the park, including the dependency on their husbands or other male or female family members, is a main theme mentioned and criticized by women when talking about accessibility to Pardis Banvan.

The limitations associated with accessing Pardis Banvan seem to be the problem women complain most about, to the extent that authorities have been forced to begin finding a solution. The park guard mentioned that the municipality had started to explore the possibility of opening another entrance at the park's northern corner close to the shuttle taxi terminal, which has not been realized to date (Summer, 2015).

Finally, accessibility requirements for those with limited mobility and mothers with baby carriages are more or less observed in Pardis Banvan. The slope's design at the park's entrance, which is meant to block its visual permeability, has been done in a way that provides a wheelchair ramp with a 1:12 slope (the standard as per the American Disabilities Act (ADA) and the Barrier Free Access (BFA) guidelines), and thus easy access to visitors using mobility aids or mothers with baby carriages. The flat design of the park also provides access to different functionalities. With the exception of sunbathing facilities, where the roof is used for sun exposure but an elevator is lacking, all service and facilities are single-story constructions and are not accessible or equipped with automatic doors and other similar accessories.

Accessibility of the women-only Pardis Park in Isfahan

The women-only Pardis Park in Isfahan is located in a partially developed part of the city surrounded by farms, private gardens and agricultural enterprises belonging to the Nazhvan district (a district known for its greenery and hiking and trekking opportunities). The park is situated on the side of a highway and a small exit allows visitors to access it. The park's location also affects its legibility and ultimately its accessibility, making it accessible by car. However, because subways and shuttle taxis are absent, buses are the only means of public transportation to the park. Nevertheless, visitors do not consider buses to be a comfortable means of transportation, partially because there is no bus stop assigned to the place (even though there is a stop a rather close distance to the park's entrance) and because visitors have to walk by the highway to reach the park. Almost all visitors observed during the study commuted by private cars and taxis. The park is also connected to the transportation network by the well-designed and well-traveled transportation infrastructure in Isfahan, and through the highway by which the park is located even though access is limited primarily to private cars. Thus, in terms of accessibility, cars are the only means of easy and comfortable access to the park, especially given that space is provided for parking free of charge.

There is also an area designated for pedestrian access to the park (mostly coming from newly-erected residence areas). However, many visitors prefer to use private cars rather than walk to the park because of security concerns and apprehension at having to walk close to noisy and crowded highways. Given the fact that this park is one out of five women-only parks in Isfahan (two of which are in the Nazhvan district in the city's west side), it is meant to serve neighborhood locals while at the same time keeping the neighborhood a fair distance from the park. Despite all limitations, the park remains accessible to the residents for whom it is meant to serve (see Figure 2.3).

Furthermore, the park is accessible without any major hindrances for those with limited mobility and mothers with baby carriages. The absence of stairs and obstructions at the park entrance provides good access to the park, while its flat design makes it amenable to good in-park accessibility. However, to provide facilities is one thing, but to have cultural and traditional norms that incorporate and encourage those with disabilities to actually use those facilities, is another. To what extent society's traditions and normative values encourage people with disabilities to take part in public spaces remains to be studied. Such an endeavor would offer worthwhile comparative data given that encouragement may be less persistent and visible in such cities as Isfahan and in families with more traditional ties.

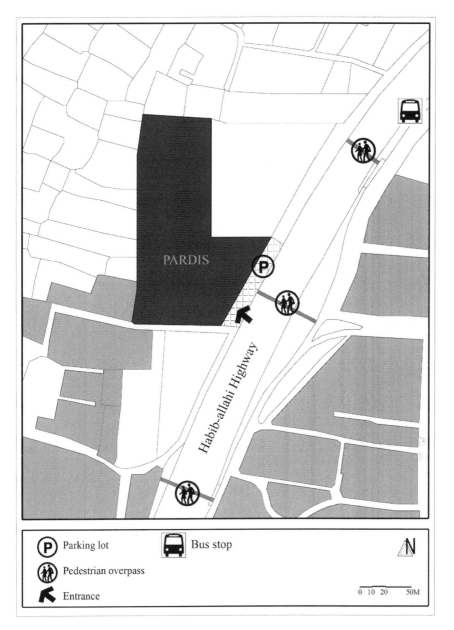

Figure 2.3 Accessibility by various means of transportation to the women-only Pardis Park in Isfahan.

Accessibility of the women-only Park Banvan in Rasht

Park Banvan is situated in the most central and densely populated area (as well as in one of the oldest parts) of downtown Rasht. Close to such cultural and educational centers as the university and the bazaar – the commercial heart of the city – the park is an excellent example of good accessibility. The park is, however, a fenced green space located at the city's heart that is solely for use by women, and lacks amenities and services compared to the three women-only parks in Tehran and Isfahan, which have been previously discussed. Despite its lack of facilities and services, women extensively use the park because of its vicinity to the university and bazaar. Also, since it lacks any entrance door, Park Banvan is permeable from all sides. Rather than doors, these entrances are symbolic gates, to mark the boundaries of the park. From the west, the park leads to the most crowded street in the city and is accessible by foot from the neighboring residential and commercial fabric.

As in many other smaller cities in Iran, taxis, with their reasonable fare, are the most popular means of public transportation and are easily available around the park (see Figure 2.4). Easy to access by foot and public transportation, those using private cars have a difficult task of finding parking near Park Banvan precisely because of its centrality. The park is accessible from two points, one entrance from the east and another one from the west, with the northern and southern sides of the park closed. Moreover, the park's openness and its flat organic design provide good in-park accessibility for users. This also makes it possible for people with disabilities and mothers with baby carriages to enter and move easily inside the park. Despite the fact that park is not used to its full capacity, partially due to the fact that other parks in the proximity are easily accessible, Park Banvan enjoys good access both city-wide and in-park.

Accessibility of the gender-mixed Niavaran Park in Tehran

Niavaran Park, located in northern Tehran on a hillside of the Alburz Mountains, enjoys the vicinity and diversity of its natural surroundings. Located close to the mountains, beautiful views and relatively clean air on the one hand, and being home to a royal palace of the same name on the other hand, has given the Niavaran neighborhood a privileged position in Tehran. This in turn has also affected the visual character of the park and its functionality. Since the Niavaran Park is located in a residential area and surrounded by various compatible functions and services, it provides good connectivity to the surrounding urban fabric and offers a variety of access to its visitors. Pasdaran Street surrounds the park from the south and west and provides the main access to it. Multiple bus stops and taxi stations at Pasdaran Street connect the park to the city's public transportation network. From the eastern border, the park is accessible by a few alleys. A parking lot, located in the south of the park, provides parking spaces for those who use private cars. In addition to all the other ways to access Niavaran Park, most park users live within walking distance from it. Insights from interviews with locals reveal that, other than some elderly people living in the neighborhood who prefer

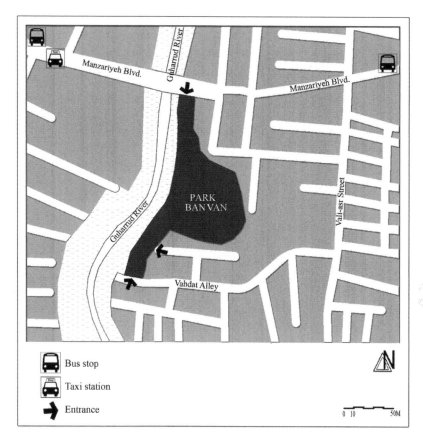

Figure 2.4 Accessibility by various means of transportation to the women-only Park Banvan in Rasht.

to ride in their own private cars or public transportation in order to avoid the steep slope of Niavaran's streets, many users prefer to walk to the park.

What has marked out Niavaran Park is that it is situated within an old and well-established neighborhood containing a combination of various compatible urban functions. The fact that a number of cultural, scientific and artistic centers are located in the park's neighborhood has contributed greatly to its accessibility. The neighborhood's higher socio-economic position and people's interest in using the park are also additional advantages.

Since the park was originally meant to serve the local neighborhood – an intention which later changed for a number of reasons discussed under "A place for all" – it was designed to provide easy access for pedestrians from different points. From the east and west, the park shares a border with the Niavaran district's residential fabric. Since some of the park's sport facilities are located on the east side, and all the smaller streets leading to the park are blocked, there is no accessibility from the east and residents living on that side have to walk to access the park either from the west or south, which do provide good access. The west side of the park houses the celebrated Niavaran Palace and Museum, which many consider a pleasant path to walk. However, while accessibility in terms of access points is possible without much difficulty, the park is located on a hill slope which requires physical exertion for many, especially senior citizens. Park users from the neighborhood who arrived at Niavaran Park by car emphasized that their homes were within a close walking distance to the park but that they preferred to drive for one reason or another. Many said that the park is "up the hill and not easy to access if you are old and suffer pain in your joints or legs." This becomes particularly problematic when the weather is cold or when the street surface is frozen.

Access to Niavaran Park by taxis and shuttles is also convenient. There are cabs and shuttles to the park from Tajrish Square (a major traffic hub in northern Tehran). As noted earlier, the park's location in the middle of a complex of compatible (scientific, cultural and artistic) functions has greatly contributed to its accessibility to the larger city-wide transportation facilities. There is a shuttle stop at the park's entrance which further facilitates access to the network of greater Tehran (see Figure 2.5).

Since Niavaran is a district with a higher socio-economic status compared to many other parts of Tehran, the culture for using private cars over public transportation is more prevalent and visible in this part of the city. While finding a place to park the car is always a challenge for this area, a large free-of-charge parking lot located close to the park's entrance (which we were told is used for parking because the property's ownership was disputed and left idle, and thus it could not be otherwise used for construction) provides easy access for visitors. In addition to the free and available parking, there is a bus stop close to the park and a public transportation service is provided for the park and neighboring facilities, which is mostly exploited by visitors from different parts of the city. Furthermore, the park is extensively used by younger (particularly male) visitors from southern Tehran who find social interaction there (not least with the opposite sex) easier than in their own residential neighborhoods, with more traditional and religious sentiments. Younger visitors reach Niavaran Park by motorbikes which come in handy in Tehran's heavy traffic, while families who come to picnic in the park arrive in private cars.

Mobility from southern Tehran to the north (and never vice versa) partially happens due to northern Tehran's attractiveness as a district with people publically considered cultured and because of its higher socio-economic status and cultural taste. This is an intriguing phenomenon that will be addressed later in this volume. The combination of the park's good accessibility and connectivity to different parts of Tehran, whether through public transportation or private means, along with its mixed-gendered space, attracts visiting singles and families from other parts of the metropolis, especially on weekends and holidays when families want to spend time together.

The stepped design of the main (northern) entrance to Niavaran Park, and its topographic characteristics, makes access a challenge for visitors with limited mobility or mothers with baby carriages. The park's design dates back to a time

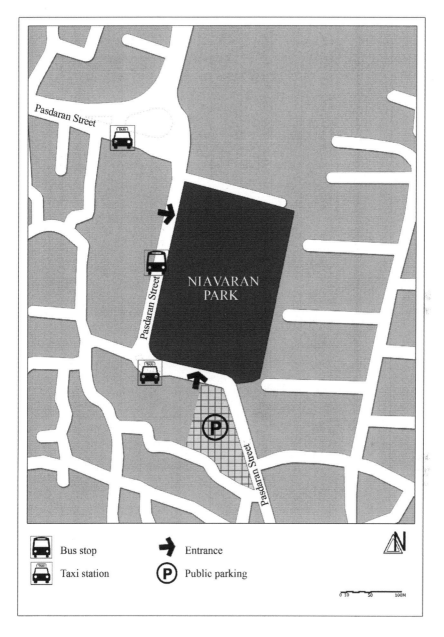

Figure 2.5 Accessibility by various means of transportation to the gender-mixed
Niavaran Park in Tehran.

when requirements taking into account this population's needs were not imposed
on the designs of public spaces. Also, the later addition of wheelchair ramps has
not been possible due to heavy costs and because parts of the park would be either

destroyed or demolished in the process. The southern entrance, however, does provide access for users with limited mobility because it is flat. Thus, the park's topography and lack of wheelchair ramps to accompany the stairs at the main entrance provides limited access to Niavaran Park except at its flat southern half.

Accessibility of the gender-mixed Bi'sat Park in Tehran

Located in southern Tehran, Bi'sat Park is well connected to the surrounding urban fabric. Bi'sat highway in the south and three other streets at different sides of the park connect it to greater Tehran's transportation network. A cross-country bus terminal (a public transportation hub serving destinations throughout the country), which is located at the park's west side, provides various public transportation facilities (subway, bus and taxi) to most parts of the city. Several bus stops close to the park's main entrances in the southern and western peripheries provide flowing access to the park from different parts of the city (see Figure 2.6).

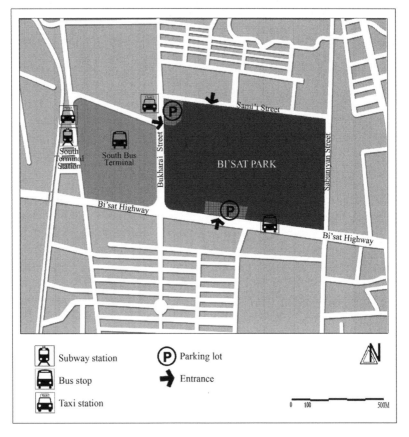

Figure 2.6 Accessibility by various means of transportation to the gender-mixed Bi'sat Park in Tehran.

The surrounding fabric – especially from the north, south and east – is also composed of larger residential complexes, which makes the park accessible to locals by short walking distances.

Bi'sat Park covers an entire block and is surrounded by streets of various widths. Although the park is permeable from different sides, the main entrances are marked on the western and southern sides. To the eastern side is Tehran's South Bus Terminal, which is connected to the entire city by subway, bus, shuttle and taxi services. Despite this connectedness, and because the destination by all these means of public transportation is the bus terminal, there is no easy route from these stops or the station to walk into the park. The first immediate block next to the park is not part of the residential fabric and pedestrians from the neighborhood need to walk through facilities, but mostly industrial storage and gross distribution hubs, to reach the park. Regardless, access to Bi'sat Park is still rather easy because the park is permeable from all sides.

To its northern border, the park neighbors a school and two universities (Azad University Tehran South Branch and a not-for-profit university) and also connects to a residential area. This is where most people living, studying and working in the neighborhood access the park. Although the park is close to the bus terminal and thus acts as an open-air waiting room for passengers traveling to and arriving from different parts of the country, and although its size has turned it into a city-wide park used extensively by different groups of people, it was originally established and is still regarded as a local park meant to serve the immediate neighborhood. Despite its changed function, the park is rather conveniently accessible for visitors from the neighborhood. The park is also easily accessible for those passengers waiting for their buses at the terminal. By simply crossing the street, the park becomes accessible from any given point. All in all, Bi'sat Park is pedestrian friendly, and despite the incompatible surrounding fabric, is easy to access by locals from neighboring areas and visitors originating from the bus terminal.

As noted earlier, the park is situated at the heart of a transportation network because of the bus terminal which connects the neighborhood to the rest of the city by multiple transportation services. To access the park by these means is easy and offers a variety of ways by which to travel to and from the park. The bus network (with its separate express services for the crowded parts of the city) efficiently connects Bi'sat Park to the rest of the city. A bus stop at the very entrance of the park further facilitates the use of buses.

Pedestrians do not favor traveling the route to the park from the subway, although it is some 250 meters of distance. The reason for this that visitors repeatedly mentioned is the incompatibility of the neighboring functions. The street to the subway station is mostly occupied by buses and their drivers who use that area for parking, for minor maintenance work and for cleaning their vehicles. Visitors also mention the absence of a shady and verdant route in the summer (along with the air pollution, which due to Tehran's geographical characteristics is concentrated in this part of the city more than in the north) and the route's exposure to breezes (partially due to lack of vegetation which is a result of polluted substances left by the buses) as other sources of inconvenience.

Although many women did not personally experience any instances of physical harassment or crime, "*feeling* uncomfortable and unsafe" was the main reason for preferring the bus network or taxis to access the park. The bus station at the park's southern corner connects it to the southern part of the metropolis (the route is to the Khavaran Bus Terminal which in turn is connected to the rest of the city). The fact that the park is solely accessible by transportation from the north and east has by no means affected its accessibility.

For those traveling by private car, there is a parking lot at both of the park's entrances and parking spaces are provided for a small fee. Park visitors can also use neighboring streets for free parking on weekends and on certain special occasions.

Despite a number of limitations, Bi'sat Park is very accessible not least due to the neighboring bus terminal. Access to park is easy, fast and safe, and there exists opportunities to choose between various transportation alternatives. Still, one question remains to be answered: whether and to what extent being gender-mixed has contributed to the park's accessibility. One of the main criteria for selecting a location for a women-only park is its invisibility from surrounding functions, and as such it is usually far from transportation facilities. In many instances, these parks are located in parts of cities that are not fully developed since this provides a better ability to control development according to the rules established for erecting a women-only park in the area. This could be observed in all women-only parks (except Bihisht Madaran which has changed from being gender-mixed to women-only). In an endeavor to speculate the feasibility of changing Bi'sat Park to a women-only park, one may realize that the location and accessibility (along with permeability as discussed in the following section) proves impossible for its conversion. Hence, the notion of accessibility is an important element for Bi'sat Park and is achieved, in part, due to its actual accessibility for all users.

The park is also used by locals as a space to take shortcuts to other areas. Many prefer to use the transportation network to reach the terminal and then walk across the park to their homes. Users repeatedly noted this as one of the park's advantages, which would be diminished if the park was a women-only park and which would eventually affect accessibility. Thus, Bi'sat Park's location, at the heart of an urban fabric with multiple functions, has tremendously contributed to its accessibility from different parts of the city.

In designing access within the park for people with disabilities, Bi'sat Park meets the requirements of a maximum 1:12 slope. This, in combination with the park's rather flat topography, has allowed accessibility for people with specific needs and mothers with baby carriages. In the few instances where the slope proves too sharp, thus forcing visitors to consider using stairs, alternative routes are provided for those with limited mobility. Furthermore, the park's organic layout has made it possible for its designers to use multiple avenues and ways to connect various parts of the park.

Accessibility in women-only parks versus gender-mixed parks: a comparison

Both of the gender-mixed parks described in this volume, Niavaran and Bi'sat, are well connected to the surrounding urban fabric as public urban spaces and meet accessibility criteria. The accessibility and connectedness among the women-only parks, however, vary greatly. Bihisht Madaran Park in Tehran and Park Banvan in Rasht, both originally gender-mixed parks converted to women-only spaces, have good accessibility through all means and transportation, including that of pedestrian travel. The two women-only parks that were originally erected as such are not easily accessible and connected, and thus satisfy criteria mandating concealment and invisibility for women-only parks. With the exception of two of the women-only parks, diverse transportation choices improve these parks' accessibility making them available to diverse visitors. All the parks serve users from their surrounding areas in the neighboring districts, but some with better connectivity (among other reasons) serve various groups beyond the neighboring districts.

A successful urban space should easily be accessible to all including the vulnerable groups such as the elderly, disabled, children and the like. Niavaran and Bi'sat gender-mixed parks could be regarded as good examples of such spaces. Located in mixed urban contexts, these parks provide easy access for a wide range of users with different needs and limitations. However, a woman visiting Pardis Banvan Park in Tehran or Pardis Park in Isfahan will face a series of constraints toward access. The accessibility criteria are better for the women-only parks of Bihisht Madaran Park in Tehran and Park Banvan in Rasht.

Functioning mainly at the neighborhood scale, most visitors to the parks under this study's scope (both women-only and gender-mixed) are likely "pedestrian users" who live at a stone's throw's distance from those spaces and can reach them with a short walk. Once again, the surrounding urban areas should be pedestrian friendly in order to encourage more people to use a park. For this to happen, a park's location is important. While women in parks located in a city's downtown or to its north (Bihisht Madaran Park, Niavaran Park in Tehran and Park Banvan in Rasht) "feel safe" walking to them, "feeling unsafe" was one of the main concerns limiting accessibility for users of the parks located in the southern part of a city or on its outskirts (Tehran's Pardis Banvan Park and Park Banvan in Isfahan).

Connectivity to the urban fabric, neighboring compatible urban functions, existence of residential blocks, small-scale businesses and other public places can increase the pedestrian flow in an area and hence provide a pedestrian-friendly environment. Street design is also crucial in lending the feeling of safety and encouraging the use of a space. With these points in mind, one may safely conclude that Bihisht Madaran women-only park and Park Banvan in Rasht, and also the mixed Niavaran Park, are located in more pedestrian-friendly environments compared to the women-only Pardis Banvan Park in Tehran and Pardis Park in Isfahan. The commercial, cultural and residential functions surrounding Bihisht Madaran Park; the bazaar and university along with a landmark river close to Park Banvan in Rasht; and a series of residential blocks in the east, a cultural center in the west, a famous bookstore and a research center in the south and a historical palace and museum in the north of Niavaran Park have all created spaces involving a mixed urban fabric with strong recreational and cultural characteristics.

The women-only Pardis Banvan Park in Tehran with its island shape and isolated location at the edge of the city; Isfahan's Pardis Park in the middle of a farming structure; and the gender-mixed Bi'sat Park, with its neighboring large-scale functions such as a bus terminal, a power station and a number of industries, all have limited walking accessibility and that has resulted in less pedestrian-friendly environments, which in turn has resulted in limited accessibility

A space's accessibility is also affected by its "permeability" with a place being considered more accessible if it is permeable. While the public urban spaces of the gender-mixed Niavaran and Bi'sat Parks, as well as the women-only Park Banvan in Rasht, are permeable and thus better accessible both physically and visually, women-only parks lack these criteria in general. In addition to their several entrances, there are no physical barriers to many parts of the mixed parks, and thus the opportunities to enter these parks from different spots increases. Such a thing does not exist in women-only parks since enclosure is a main criterion for these spaces. In other words, only the mixed parks (and to certain extent Park Banvan in Rasht) meet the criteria for the physical permeability defined as a basic requirement for these parks. The mixed parks are also separated from the surrounding urban fabric by rows of trees or short brick walls, which while delineating borders, still provide visual connectivity and still expose the parks' exteriors to visitors. Having the parks' exteriors visibly open also allows visual access to those who pass by the parks, giving them opportunities to glance into the inside space.

Allowing people to see a park from the outside or "visual permeability," also known as "visual connectivity," plays a major role in inviting people to use the space and, hence, contributes to a park's vitality. People can see a place before entering it and then consider if they would feel comfortable, welcome and safe there. This visibility also provides possibilities for soft control and informal surveillance over a space, which is mostly carried out by citizens rather than the charged authorities. In fact, existence of hard, tall and impermeable walls surrounding the women-only parks hinders those possibilities, which are considered fundamental requirements for such public spaces as urban parks.

However, despite permeability's importance in creating safer (by means of visibility) and welcoming urban spaces, it would be a mistake to only consider design elements in this process while overlooking the social and functional factors at play. Many women noted, "if walls provide women with a park of their own, a space to give them a sense of freedom and dignity," they welcome it. Furthermore, comparisons between the women-only Bihisht Madaran and gender-mixed Niavaran parks on the one hand, and Pardis Banvan and Bi'sat parks on the other, due to contextual similarities reveal the limitations in solely looking at these parks' technical concerns without considering social values and norms. Hence the technical aspects of permeability must be studied along with the social settings in which the parks are located. Due to the layout and accessibility from four streets at different sides of the park, technically speaking permeability is stronger at Bi'sat Park compared to Bihisht Madaran. However, Bihisht Madaran and Niavaran Parks' social surroundings and compatible functions have compensated for the technical limitations (imposed by the walls for Bihisht Madaran and created by the palace complex located at its northern edge for Niavaran Park), thus making them safer, more pedestrian friendly, attractive and used by the public.

Legibility

Legibility (recognizability) aims to improve peoples' perception, understanding, experience and enjoyment of a place. Integrating information, identity, and both physical and functional dimensions of a legible environment help create a comprehensive and smooth movement for park users between various destinations. "The degree of choice offered by a place depends on how legible it is: how easily people can understand its layout" (Bentley *et al.* 1985). Carmona (2003), in an argument against Lynch (1960), the American urban planner who assumes legibility as a "secondary problem for most people," notes that the majority of urban designers acknowledge legibility as one of the first necessary features in creating a successful public place. They further assert that legibility should be taken into consideration from the very initial stages in the design process.

For a women-only park, legibility has a two-fold significance: city-wide legibility in a larger-scale context, and in-park legibility on a much smaller level. Any given space's location, its connectivity to other functions and its physical characteristics are factors that could contribute to a more legible public place on a city-wide scale. For instance, Bihisht Madaran Park located on natural hills in a central part of Tehran and alongside one of the city's main highways, is more legible than Pardis Park in Isfahan which is located in the suburbs and in an area unfamiliar to many of the city's inhabitants. As one of the park's visitors remarked, "I could hardly find the park, even though I knew where it was located and even after asking many people." Despite the visibility of Bihisht Madaran some 53.5 percent of women in a study conducted in the park's neighborhood note that they have only heard about the park through word of mouth (Kawsari 2008: 17).

A public place's size is also an important element for its legibility. For instance, having an area of 1.8 hectares, Pardis Park in Isfahan makes it less legible than Pardis Banvan Park with its area of 27 hectares. Additionally, the physical and visual characteristics of a place can also facilitate locating it more easily. A large row of tall trees behind a wall, for example, could represent the park for someone viewing them from a distance.

Upon arriving at the park, the scale for a visitor shifts from the larger city context to the specificity of the place. A well-designed environment should support its legibility by using typical functional characteristics. Legibility, according to Steiner and Butler (2007: 18), is about providing recognizable routes, intersections, landmarks and anything attractive to help people navigate inside a park. It should reinforce a sense of place by bestowing visitors of a park with the ability to locate important design elements and then apply them to find their routes.

In a place that is not legible enough, people tend to become lost, and the likelihood of losing direction increases with the place's size. Using landmarks, signs and symbols, and providing information to guide visitors in different parts of a space are vital. Many women-only park users who were interviewed in this volume repeatedly pointed out the difficulty of locating a given facility in a large park. They note that there are not enough signs in these spaces to show the right directions.

Legibility can also be influenced by a sense of place. People remember a place more readily by addressing memorable components, unique experiences, physical monuments or landmarks attributed to it. This sense of place helps people to create their own image of a space and to navigate subsequent visits better. Instances of such features that visitors could use to locate or navigate park spaces include for instance the statue of a mother figure in Bihisht Madaran or the lake in Pardis Banvan Park.

There are also various methods of helping people improve their perception of a space. Modeling and mapping are among the most practical methods applied to large-scale areas. Since the spatial layout for large public places like parks cannot be immediately perceived, a map, a guidebook or a model can help visitors with locating their whereabouts and with spotting places they need much easier than if they did not have these tools. Maps and models also provide an overview of the existing places, amenities and services located inside the park. Visitors to the parks interviewed in this volume repeatedly complained about the lack of information signs in these places; a complaint that was confirmed through the researchers' observations and through comparing different parks.

To better comprehend and assess the notion of legibility through empirical methods, a number of observations were conducted in each park under this study's consideration. The result of these observations has aided in identifying specific and detailed features concerning legibility in each of these parks, and at both the macro (city-wide) and micro (in-park) levels.

Legibility in the women-only Bihisht Madaran Park in Tehran

Bihisht Madaran Park's large size (19 hectares) and its location at the top of the Abbas Abad Hills surrounded by three main highways (Risalat, Haqani and Himmat), make the park rather easy to locate with city maps and guides so as to plan a visit or to find ways of commuting or accessing it. Situating the park within its urban context via the internet and electronic media is also simple. The Abbas Abad district, in which the park is located, is within a very well-connected central area of Tehran, and also in a well-established part of the city that is appreciated for both its commercial (including cultural) and residential functions. The walls along these highways are covered by large prefabricated cement above which Bihisht Madaran Park's trees are visible, thus enhancing the park's overall visibility for visitors arriving from the highways. Although the park's general appearance and ultimately its permeability are affected by the restrictions required for women-only institutions within the Iranian context, the abundance of longstanding greenery and these tall trees stretching above the walls make the park legible and recognizable within the urban context. Despite drawbacks caused by limited permeability, visitors receive the discernible sense of approaching a park when arriving from the surroundings highways.

The park's location at the top of a hill also greatly contributes to its recognizability. The abundance of directional signs and guides along the way to Bihisht Madaran Park helps visitors find their way to it without major complications. Moreover, the park is further legible since it occupies an entire block and it is rather easy to navigate the perimeters to find its entrances. Against this description, one would expect visitors to distinguish the park from a distance; however, since the park is surrounded by tall commercial and residential buildings, it becomes more recognizable when close enough to see its enclosing walls.

Most likely, Bihisht Madaran Park's visibility would improve if its permeability were increased. A potential visitor (except when arriving from the south) cannot simply spot the park from a distance and then drive toward it. In the respective urban-study literature, the city's sporadic expansion in the absence of a comprehensive urban plan is discussed among the main reasons for such obscurity. The constraints in the park's visibility for first-time visitors were brought up by women, especially by those who were new to the area.

Despite its invisibility, Bihisht Madaran Park was known to people living in its neighborhood before it became a women-only park, and later city-wide through Tehran Municipality campaigns. The park's well-established reputation within its urban context has contributed to its city-wide legibility (though not visually).

Bihisht Madaran Park has two specific features which create variations to in-park legibility: the area's slope and the park's organic (non-symmetrical) design (see Figure 2.7). For the first-time visitor, these two characteristics create difficulty for forming an impression of the park, in grasping its overall image and in reading the totality within it, including that of its different facilities. This limitation is compensated to a great extent by a three-dimensional scale model (located at the northern entrance, known as the main entrance), maps (at all entrances) and signs for direction.

While the park's design seems organic, examining the three-dimensional scale model or maps permits an understanding that all the park's avenues lead to a central point in its middle. There, a statue of a mother figure has become a landmark with which navigation inside the park is facilitated. Although organic and non-symmetrical, the park's design follows a certain logic stemming from the limitations imposed by the topographic characteristics of the park's venue.

Legibility in the women-only Pardis Banvan Park in Tehran

Since Pardis Banvan Park is an island in the midst of a network of freeways, and because it is located at the very edge of southern Tehran's urban fabric, the park is out of everyday reach. Unless a visit to the park is specifically planned, one's way never crosses the park. Essentially, the entire district in which the park is located lacks attractions, whether cultural or otherwise, to draw visitors from other parts

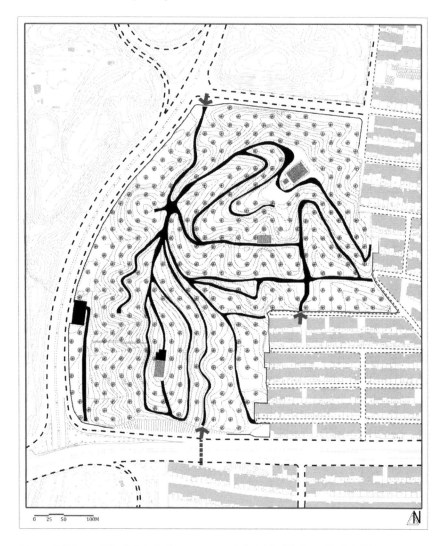

Figure 2.7 In-park legibility in the women-only Bihisht Madaran Park in Tehran.

of the city. Though this is regarded as a disadvantage for such public space as parks, it was the main advantage attracting the attention of urban planners. The venue was separated from the rest of the urban fabric and hence it was easier to create a women-only space.

The surrounding freeways, which serve as the main routes connecting Tehran to the southern part of the country, do not necessarily provide connectivity to Pardis Banvan Park. This is partially due to the park's legibility problem. As noted earlier, a park's age, overgrown vegetation and greenery provide good readability in an urban context. As a newly established park, Pardis Banvan lacks these advantages. Because trees in Pardis Banvan are newly planted and thus not tall enough to see from afar, as in Bihisht Madaran Park, they do not give the impression that the space is a park covering 27 hectares. It will take a long time for these newly planted trees to grow and become visible from behind the park's tall brick walls. Even then, the park area may resemble more of a factory or some type of institutional space rather than that of a park. Furthermore, the park's brick walls, with their symmetrical patterns, neither give off the aura of a park nor that of a leisure or entertainment space.

When using maps or electronic services found city-wide, the park is easily discernible. This is a result of the park's size on the one hand and of its distinction from the rest of the district's functions on the other hand. The problem with the park's legibility and visibility has been greatly aggravated by the lack of signage for it in the surrounding area, and by its single entrance. In fact, unless one knows where and how to reach the park, it is almost impossible to find it despite its size. The entrance itself, however, is very well designed. There, a large sign invites visitors into the park, while the entrance portal conceals the park's inner space giving the feeling that one is entering into a gated space.

Pardis Banvan Park's vicinity to two other larger parks (one male-only and the other gender-mixed) leaves the visitor wondering if these parks are all used to their full capacities. Because the space between these two parks was abandoned and isolated, which met the criteria set for women-only spaces, it seems that it offered a unique opportunity to be transformed into a women's park. Paradoxically, however, the proximity of the three parks has provided visitors with new dynamics for using them. For example, during field visits, families were observed on a day outing where women were dropped off at the Pardis Banvan's entrance, but men proceeded to male-only water sports Azadigan Park. Later, the family reunited in the mixed park.

In-park legibility is achieved through Pardis Banvan Park's discrete symmetrical design, which is inspired by the Persian *charbagh*, whereby two axes cross providing for four separate spaces. Each axis leads to one of the main facilities and creates a functional landmark. The park's only entrance opens to the middle of the longer (horizontal) axis and, despite the fact that a visitor is unable to see the park in its entirety upon arrival (not least because of its size), its symmetrical design provides good overall legibility (see Figure 2.8). A large map located at the park's entrance also facilitates legibility. The vegetation and greenery's lesser degree of density also adds to in-park legibility and visibility.

Despite all of this, the park's size and lack of interior signage for directions, results in services or facilities being far from hassle free. As one young visitor puts it,

> When we finally reached the park, we were really thirsty. My friend and I decided to grab something to drink before starting our excursion, but finding a vender to buy water or juice became an excursion in itself.

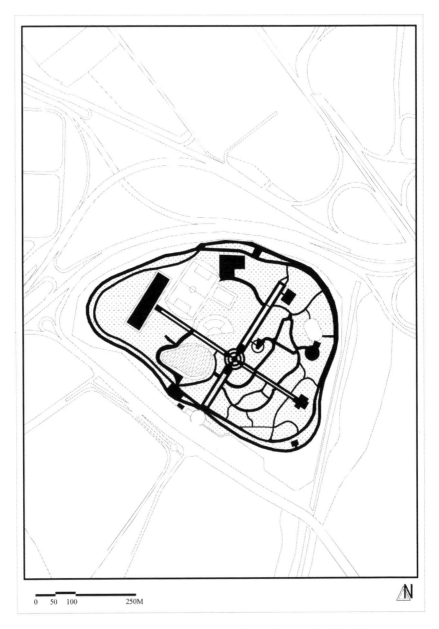

0   50   100   250M

Figure 2.8 In-park legibility in the women-only Pardis Banvan Park in Tehran.

Legibility in the women-only Pardis Park in Isfahan

Pardis Park in Isfahan was built on the city's outskirts; the area is not yet developed and is not at the crossroad to any major urban function. This, along with the park's small size, has greatly affected legibility in its urban context. The park is surrounded by brick walls. However, unlike Pardis Banvan in Tehran where the walls give the impression of a well-maintained institution or public function of some kind, rather than a private farm or garden, walls around the Pardis Park in Isfahan are built with low quality materials and less care. The walls seem like they were a cheap undertaking and lack the visual attractiveness that invites a passerby or visitor inside the park. Even though tall trees extend the walls, they look disconcerting because of the sharp steel fences mounted on top of the walls which surround the park. There is no sign to guide the potential visitor to the park, except when reaching the actual entrance. Aligned with all this, one can conclude that the park has very low legibility in the urban context. This is confirmed in interviews where women say that they have heard about the park only by word of mouth from previous visitors or friends.

The park's small size and its location in the middle of farms and private gardens on the city's outskirts, has made it difficult to find on the city map or through electronic search engines. While the size of the park has made it illegible in the urban context, it has contributed to its in-park legibility. With an area of 1.8 hectares, the entire park can easily be read upon arrival. A plane (flat surface) with sporadic vegetation (due to the climatic characteristics) and lack of facilities, except a basic small hygiene facility, have also left the park with open visible space that can be viewed and read without a problem. And even with good in-park legibility, an informational sign helps visitors to locate limited services upon their entrance.

Legibility in the women-only Park Banvan in Rasht

Park Banvan's location in the heart of the commercial part of Rasht city center, but also at the bank of the river – a natural attraction that passes through the city – has increased the park's legibility. The park is very small in size and is in the middle of residential and commercial complexes nestled within the old fabric of the city. This, along with the organic, non-symmetrical shape of the park, limits legibility (see Figure 2.9). To a large extent, this limitation is balanced with the park's location, a place where people pass by almost on a daily basis to run errands because of its vicinity to a number of urban functions.

There are no signs or guides to lead visitors to the park from differing parts of the city. The only sign is at the entrance, which gives the name of the park and warns men against entering the park. In exploring the park's legibility, one should be conscious about the differences between large cities such as Tehran and smaller ones such as Isfahan or Rasht. While city-wide legibility in Tehran is dependent upon publicity and visibility within the urban context, private networking and word of mouth contribute to legibility in smaller cities, especially for spaces located at the city center. Like the park in Isfahan, the smaller size, high permeability and lack of facilities and services have contributed to better in-park legibility. Finally, the climatic characteristics of Rasht, with its dense vegetation and flora, rival the parks and, despite good legibility, have made it difficult to read Park Banvan in the urban context for people who are unfamiliar with the city.

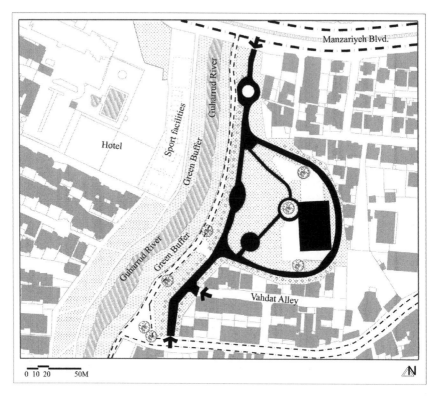

Figure 2.9 In-park legibility in the women-only Park Banvan in Rasht.

Legibility in the gender-mixed Niavaran Park in Tehran

Originally established as a local park to serve the surrounding neighborhood, Niavaran Park has changed over time, not least due to the sharp north–south dichotomy drawn within the city of Tehran. Its location, as well as being part of a recreation complex with a palace museum and cultural center, has turned the park into a city-wide attraction, thus rendering it legible within Tehran's urban fabric. The park is not only legible in maps, media and city promotional materials, but also to a newcomer arriving in the Niavaran district since it is visible from a long distance.

The park has also turned into a landmark, to the extent that one can use the park (and/or the Niavaran complex) as an address of its own. Due to a series of historical references, the park exists in Tehrani inhabitants' social psychology. The park's presence in people's mental map and the relative openness of the district, not least because of its social distinction (in the Bourdieuian sense), has made Niavaran Park very legible. These traits have made the park attractive for visitors from other parts of the city with lower socio-economic backgrounds.

Physically, a row of older, tall plane trees located on the southern and western sides of the park (along Pasdaran Street) introduces the park to visitors. The visual and physical permeability, and the visual connection between both the park's exterior and its interior, gives the sense of accessibility and welcomeness to people passing by and potential visitors.

The *charbagh* model of the Persian garden constructed as the park's main design, along with its location on sloped land, provides unique in-park legibility. Entering Niavaran Park from the main northern entrance, one finds oneself on top of a hill at the end of the design's longer axis, which crosses over the park. Water flowing through this axis to a main basin at the park's center (which is yet another Persian feature), as well as the park's non-organic design and its geometrical division, bestows a sense of spatial visual readability when entering the park. Also, the main entrance is located at the park's highest point and, while the panoramic view gives a holistic picture of the space, one reads the park's details once descending toward its different parts.

In Niavaran Park's geometrical plan, encompassing two perpendicular axes inspired heavily by the *charbagh* model, the axial pathway ends at a smaller water basin, while the longer axis ends at a large pool that is visible from different parts of the park. The pool functions as a powerful physical landmark inside the park, which helps visitors navigate the park and find their way within the space. The park's linear form shaped alongside the main axis, its physical landmarks, its activity nodes and the visual connectivity to the street (in the western edge of the park) help visitors form mental maps with which to perceive and reconstruct the overall layout of the place (see Figure 2.10). Because Niavaran Park is a multi-level construction, various parts of the park enjoy good visibility from different directions. This lends a sense of direction to visitors so that they do not feel lost or confused in the park.

Legibility in the gender-mixed Bi'sat Park in Tehran

Bi'sat Park was known as Khazaneh Park among the locals, but its name was changed after the revolution as part of the Islamic government's policy of Islamization concerning the lifestyle in Iran. Despite 34 years of history passing from the name change, some people still call the park by its old name. This has created an interruption in the park's city-wide legibility. Many people do not associate Bi'sat Park with Khazaneh Park, and others are confused not knowing which park a person is referring to. Otherwise put: the park's change of name has disturbed people's mental map and has confused them about the park's identity as well as about its connection to the history of the space. Regardless of the name change, the park is legible within the urban context because of its proximity to the cross-country bus terminal. The park's size (53 hectares, which covers an entire urban block), its borders shared with a number of urban functions, its history dating back some 50 years and its tall trees visible from a distance in the midst of an otherwise gray industrial area have affected positively Bi'sat Park's legibility. The lower brick wall around the park has made it visually and physically permeable,

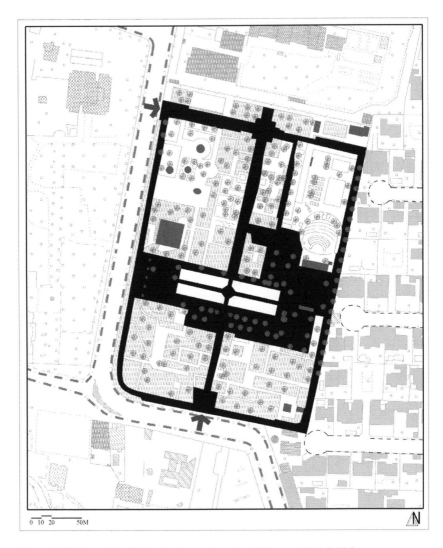

Figure 2.10 In-park legibility in the gender-mixed Niavaran Park in Tehran.

and creates constant visual connectedness between the park and the outer space to enhance legibility within the wider urban context.

Contrary to the more geometrical and symmetrical design of Niavaran Park, Bi'sat Park's naturalistic and organically designed landscape has resulted in less legibility on the micro level (in-park). Legibility is all about providing recognizable routes, intersections, landmarks and anything attractive to help users orient within a space. Bi'sat Park's curved pathways, lined with tall trees and vast green areas on both sides, have created a similar layout in many parts of the park. To a great extent, the park's large spaces, likeness of pathways and lack of attractive landmarks have bestowed a monotonous view and reduced the legibility of Bi'sat Park. Specific facilities and elements inside the space, however, assist users in recognizing various parts of the park which otherwise would be hard to navigate. Nevertheless, there is a risk that visitors may become disoriented trying to locate a specific space or facility within the park. Without prior knowledge about Bi'sat Park or using informational panels, which are sporadically provided in the park, it is not easy to navigate in the park (see Figure 2.11).

In-park legibility is also affected by the size of a given space. Bi'sat Park's enormous size, along with its organic design, has limited in-park legibility and required designers to compensate visual legibility with other methods such as landmarks (the lake, children's playground and a bridge) and signs. There is still a need for more monuments and landmarks to make in-park legibility more possible and efficient, a point that a number of visitors to the park brought up, especially those who were not regular users but who used the park while waiting to catch buses from the neighboring bus terminal.

Legibility: is it different in women-only parks?

Structural similarities between the organically-constructed women-only Bihisht Madaran Park in Tehran, and Park Banvan in Rasht contrasted against the gender-

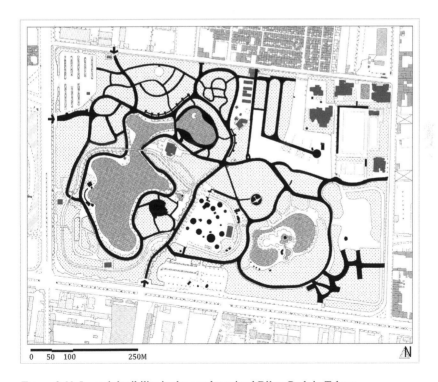

0   50   100   250M

N

Figure 2.11 In-park legibility in the gender-mixed Bi'sat Park in Tehran.

mixed Niavaran Park and symmetrically built women-only Pardis Banvan Park in Tehran, and Pardis Park in Isfahan and gender-mixed Bi'sat Park provide good grounds for comparing legibility between the parks. It also helps understand if gender-segregation policies for these parks have affected their legibility.

Legibility, with its aim to improve peoples' perceptions and understandings of a given space, affects and is affected by the physical and functional structures and by any form of information about a place. People tend to exploit memorable elements, places and activities to form their own mental maps of a place and to navigate through a particular space. These tendencies are known in urban planning and environmental psychology as "imageability" and "wayfinding" and are recognized among the most pivotal factors for making a space legible in the surrounding urban fabric at macro (city-wide) scale. Imageability, a term introduced by Lynch (1960), is the quality of an urban area that lends an observer a strong and vivid image of the space. A highly imageable city would be well formed, instantly recognizable to the regular inhabitant and would contain very distinct parts. According to Lynch, various elements of a built environment (such as paths, edges, nodes, landmarks and districts) contribute to imageability if they are appropriately placed, are well designed, are deemed meaningful and are distinct. These elements increase people's ability to see and remember patterns, and thus contribute to easier navigation and "wayfinding" within the city.

The parks' borders are marked through the usage of large, transparent and non-solid materials (i.e. bushes, trees, etc.). This maintains the visual connectivity with the parks' inner spaces, and has made gender-mixed Niavaran and Bi'sat Parks "imageable" spaces, an important characteristic that is absent in women-only parks (except for Rasht to a certain extent). It is impossible for a visitor to imagine women-only parks without entering them. Hence the interaction between inner and outer spaces – considered important for any park – is completely missing in women-only parks.

Due to other restrictions, such as a single entrance for a park of 27 hectares, one may never acquire an image of a park even when inside it. Also, the size and the high contrast between the green interior and gray surroundings, and the proximity to crowded streets with different facilities and services, have granted greater legibility to women-only Bihisht Madaran Park, Park Banvan and the mixed Bi'sat Park. The lack of any solid construction and the use of trees and bushes at the mixed parks' borders make the space imageable within the urban fabric. These also make it a traceable space for park users from a rather far distance.

One can safely argue that there are a series of methods and techniques (not necessarily complicated or costly) at park designers' and administrators' disposal that could improve the legibility of the parks. However, one can immediately recognize that not enough attention is given to using such methods. One can also see that the simple means of providing such legibility (like a more efficient distribution of facilities and services) is missing, and that certain approaches which would facilitate navigation throughout the parks are neglected. The provision of guides, maps or three-dimensional models close to the main entrances and in a location visible to visitors is used to help people learn about existing amenities and services at the parks These are also expected to provide a better understanding of the place, which is otherwise limited due to conditions and restrictions by legal and religious norms.

Enclosure and permeability

The *andaruni*/*biruni* binary division reflected in Iranian classic architecture has resulted in a distinct dichotomy of public verses private spaces in the Iranian traditional lifestyle. Whereas public *biruni* spaces, where social interactions and urban life take place, are regarded as masculine space, private *andaruni* spaces, which can be controlled and enclosed, are part of the feminine domain. While enclosure is not an alien concept in urban design, cultural and normative values in Iran uphold it as one of the unique requirements for gendered public spaces in the country.

In urban studies, "enclosure" refers to "separating different spaces" and includes defining a boundary between various spaces with different extents of privacy and/or publicness. Enclosing a space could be achieved by applying different forms and materials like walls, fences, railings, gates, arches, signage, paving, vegetation and the like, and could also be a response to "needs of safety and privacy" (Gehl 2011: 79).

Women-only parks in Iran are instances of spaces in which enclosure plays a pivotal role. These spaces, while maintained as public spaces, are required to be kept invisible from the outside. Hence, designers are forced to enclose the women-only park space and reduce visual connectivity between the outside and the inside. Observations made on the different women-only parks included in this volume show a variety of methods and materials used by the designers to achieve enclosure.

As the main feature of women-only parks compared to gender-mixed parks, enclosure meets Islamic law's requirement to close off means of exposure to women-only spaces while at the same time realizing gender-segregation policies of the Islamic administration. To maintain enclosure has been the largest challenge in women-only parks and, with the exception of the park in Rasht, it has been strictly implemented in all the parks studied here. For these parks, an enclosing wall becomes the borderline between on outer space (*biruni*) and an inner space (*andaruni*) on a larger scale. The notion of enclosure has been addressed in relation to:

- the methods of enclosure and the type of materials used for the purpose of enclosing a space;
- how enclosure affects permeability (a park as an open public space is expected to be both visually and physically permeable); and
- how it affects entrances and defines points that permeate the space.

These all are conditioned by the topographic and demographic characteristics of the venue where parks are located. In some instances, enclosure is reached by using natural barriers, while in other cases there has been a need to construct walls to separate the inner space from the outer one.

While permeability (both visual and physical) is regarded as one of the main characteristics of public parks, and is discussed extensively in urban design literature, preventing permeability is the main distinctive feature for women-only parks. The restricted permeability, however, has not been favored by the park users. Some 81 percent of the population of a study (Tehran Municipality Office of Social and Cultural Research 2011) conducted on park users in Tehran note that "permeability of a park from different directions" is a "very important or important" factor for a park as a public space. Women-only parks use restricted physical permeability (rather than unhindered) through limited entrances so as to control the flow of visitors and to solely allow women to enter the parks. Visual impermeability is a requirement aligned with the rules of religious norms, and which is strictly observed to the extent that those parks remain visually impermeable even at their entrances. There are various layers of barriers to achieve this. In most cases, a narrower pathway, similar to *dalan* [1] (in classic Persian architecture), is introduced as a buffer zone between the outer and inner spaces. Women are required to observe their *hijab* until they reach the end of the pathway where they are then permitted to remove them.

Enclosure at the women-only Bihisht Madaran Park in Tehran

Bihisht Madaran Park is an example of a women-only park where enclosure is maintained by utilizing several methods and materials in its different borders to produce a visually diverse, yet closed-off space. Some parts of the park are enclosed by trees and bushes, which not only lend a pleasant feeling to the space, but which also function as a natural insulator against noise from the neighboring highways. Soft green edges replace solid hard walls to block the view toward the park. In other parts of the park, where there is a lack of dense and high vegetation, painted surfaces or concrete walls are used to enclose the park (see Figures 2.12 and 2.13). The topography of the land, however, gives the women inside the park a panoramic view of the entire metropolis, while remaining invisible.

Bihisht Madaran Park's topography has been one of the main reasons for selecting it as an appropriate contender for a women-only park. The park is located at a slope and is surrounded by highways from three sides, which has resulted in a natural enclosure that blocks visual permeability into the park. The sides bordering the highways are enclosed using tall walls of prefabricated fiberglass plates. The walls are painted to make them look visually pleasing. On the higher levels of the slope, however, the softer barriers provided by the dense vegetation close the visual connection to the park. The natural soft barriers are reinforced by the use of steel net fences inside the park in order to make physical permeability impossible. From outside the park, these steel net fences are unrecognizable and do not disturb the aesthetics created by the green natural barriers.

On other sides of the park where there are ground level differences, the slope has offered a natural barrier and a buffer zone between the borders of the park and

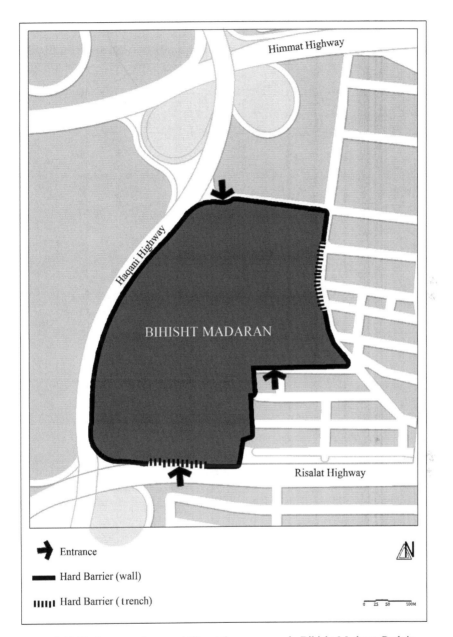

Figure 2.12 Enclosure and permeability at the women-only Bihisht Madaran Park in
Tehran.

the exterior space, thus preventing permeability. Although aesthetics has always
been compromised to achieve isolation and segregation, endeavors have been
bestowed to make the walls as visually attractive as possible. This, however, has

Figure 2.13 Enclosure at the women-only Bihisht Madaran Park in Tehran.

not always been possible, and in certain parts of the walls, especially where two methods of enclosure are used simultaneously (e.g. where the natural slope is enforced using cement walls), does not look pleasant given that parks are expected to escalate a sense of harmony and closeness to nature. The walls also function to insulate noise from neighboring highways.

The buffer zone between Bihisht Madaran Park's exterior and interior spaces is achieved with the natural slope at the entrance. The curvy paths leading to the park entrance, along with the use of stairs, have provided a *dalan* preventing visual connection with the inside.

Enclosure at the women-only Pardis Banvan Park in Tehran and Pardis Park in Isfahan

The methods of enclosure applied mainly, however not solely, in Pardis Banvan Park (but also in Pardis Park in Isfahan) give the impression that creating a private space within the larger public surrounding was the main intention of the designers of these parks (see Figures 2.14 and 2.15, cf. Figures 2.16 and 2.17). Other features such as visual attractiveness and functional convenience have been of secondary importance. This does not go unnoticed by park users. A young woman in Pardis Banvan Park reflects,

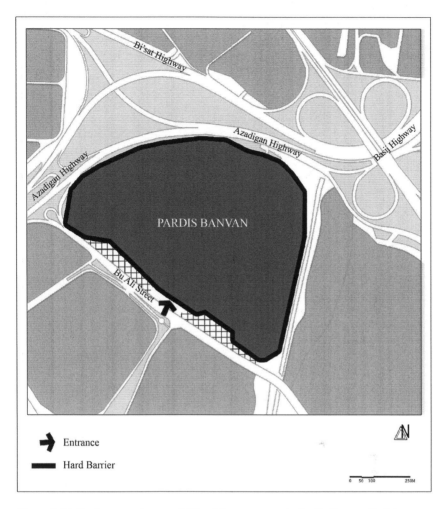

Figure 2.14 Enclosure and permeability at the women-only Pardis Banvan Park in Tehran.

Despite all, they would need a wall to close the space, I suppose. More or less like a protected swimming area in northern Iran [the public shores of the Caspian Sea, where part of the shore is enclosed and designated to women]. They have applied the same method here. I think we are used to that by now.

Another woman with a more critical tone in Isfahan notes,

That extra protection on the top of the walls is killing me! [Laughing and pointing at the knife-shaped sharp fences above the walls.] I'm totally unable

Figure 2.15 Enclosure at the women-only Pardis Banvan Park in Tehran.

to understand what's that for? It makes me feel that men are trying to invade the park and they are defending us from those predators!

Enclosure at the women-only Park Banvan in Rasht

The women-only Park Banvan in Rasht is only enclosed by a net fence, which provides visual connectivity both from inside and outside the park (see figure 2.18). Due to the visibility, women at the park are expected to observe the Islamic dress code (*hijab*) and remain veiled while using the park.

Enclosure at gender-mixed parks

Enclosure entails separating different spaces by defining a boundary between them. The necessity of such boundaries might well be dependent on various forms, functions or other characteristics of respective spaces. In the gender-mixed Niavaran and Bi'sat Parks, enclosure is used as a boundary to distinguish and differentiate the inner space of the park and outer surroundings. Contrary to the women-only parks where enclosure is regarded as a prime criterion and enforced in the form of a hard and impermeable enclosure, the mixed urban spaces of Niavaran and Bi'sat Parks are separated from outer areas by soft green hedges. This, in turn signifies that, contrary to women-only parks, the notion of privacy and invisibility of a park's inner spaces is not a significant factor

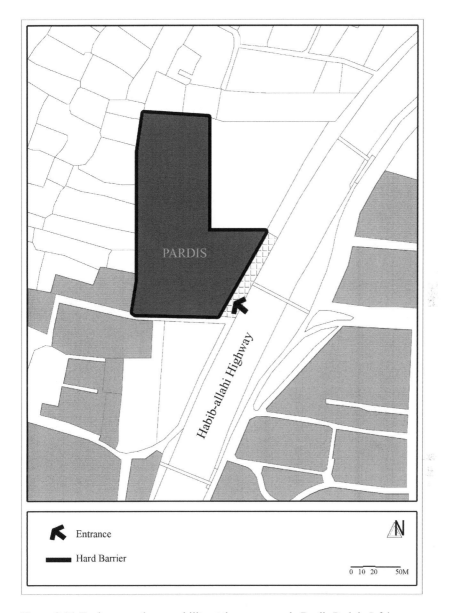

Figure 2.16 Enclosure and permeability at the women-only Pardis Park in Isfahan.

in creating mixed parks. Hence, visibility of a place from the outside connects the internal and external spaces and is likely to invite more visitors to the park, which provides greater vitality and ultimately contributes to a safer environment inside the park.

Figure 2.17 Enclosure at the women-only Pardis Park in Isfahan.

Enclosure at the gender-mixed Niavaran Park in Tehran

As noted earlier, various forms and materials could be employed to enclose a space. Niavaran Park shares an edge with Pasdaran Street, where the park's internal space in its southern and western borders is separated from sidewalks by rows of tall trees. This separation defines the park's borders while keeping visual connection between inner and outer spaces. Similar to the southern edge, the eastern part of the park, which neighbors a residential fabric, is surrounded by rows of trees used as a soft border for the park. However, due to the nature of the materials used, the border is not completely closed and provides visual and physical access to the park (see Figures 2.19 and 2.20). From the northern part, the park neighbors the Niavaran Palace and Museum, and the walls in between these two structures define the park's northern border.

Enclosure at the gender-mixed Bi'sat Park in Tehran

Compared to Niavaran Park, the enclosure of space in Bi'sat Park is done in a rather different fashion. A combination of shorter (hard) brick walls and not-so-dense green soft barriers are used to define the borders of the park and the surrounding streets. While this provides stronger visual connectedness between

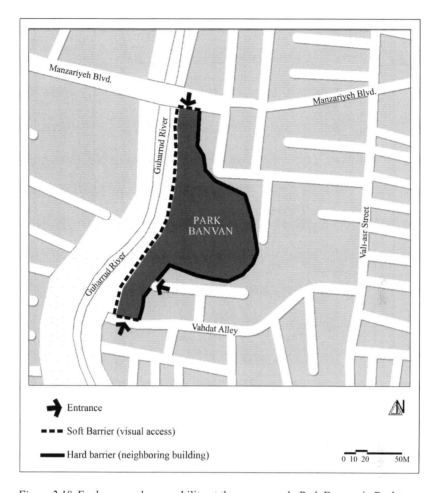

Figure 2.18 Enclosure and permeability at the women-only Park Banvan in Rasht.

the park's interior and its exterior (see Figures 2.21 and 2.22), some visitors complain about disturbing noise from the streets.

Due to the high level of noise pollution at its borders, most of the facilities and services were moved and are now located in the park's central areas. Many park users avoid the park's edges and prefer its inner parts so as to avoid exposure to noise and air pollution.

Visual attractiveness

Regardless of what goals and objectives are defined for women-only parks in Iran, they are considered special types of urban spaces. The plan and design of such parks, thus, should not only meet the quality required for a park in general, but

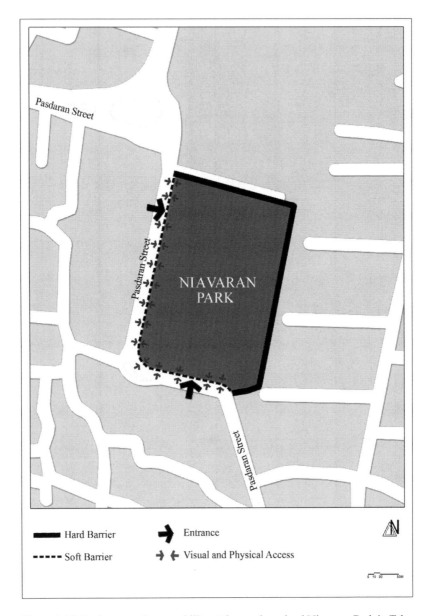

Figure 2.19 Enclosure and permeability at the gender-mixed Niavaran Park in Tehran.

should also correspond to park users' specific needs and the government's policy guidelines.

People's perception of an urban park could significantly affect whether or not they would choose to use it. Such perception is often created based on a personal ideal image of the place. Thus, to offer an image of a place close to that of an ideal

Figure 2.20 Enclosure at the gender-mixed Niavaran Park in Tehran.

image is likely to affect people's preference for using it. The visual attractiveness of urban green spaces including parks is one of the key factors in shaping the ideal image. Natural and artificial structures like buildings (of various styles), soft landscaping and greenery, furniture, lighting, signage and symbolic landmarks are among the factors contributing to the image of a park.

Parks have always been recognized as major players in the physical and aesthetic quality of urban neighborhoods. They are perceived as microclimates, what Spirn (1984: 311) calls, "bringing a piece of nature to the city" and are considered visual and social assets for communities. Urban parks improve the visual attractiveness of neighborhoods, encourage recreational and public activities, provide greater urban vitality and ultimately contribute to improving the quality of life within the city. In such a metropolitan area as Tehran, people are always in search of delightful, healthy and peaceful places for refuge from the stressful, crowded, polluted and noisy life of a hectic, large city.

Moreover, the Iranian population's demographic composition has experienced a dramatic shift in recent decades. Rural to urban migration has changed the face of both rural areas as well as cities. According to the 1956 national census, some 32 percent of the Iranian population was living in urban areas at the time. This figure, however, increased to some 73.5 percent of total population in 2015 (Central Intelligence Agency 2015). Alongside a series of consequences of this

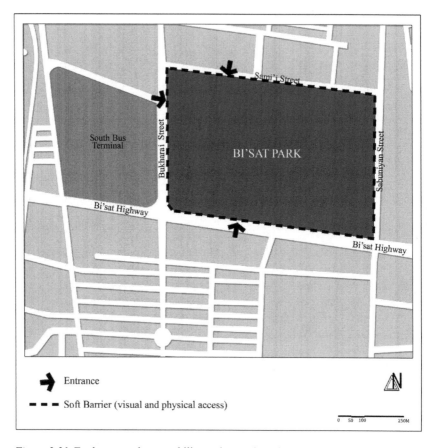

Figure 2.21 Enclosure and permeability at the gender-mixed Bi'sat Park in Tehran.

movement, perhaps the most visible was adapting to a modern lifestyle of living in block apartments with limited or no access to courtyards – a necessary component of Persian traditional houses. In the absence of courtyards with their garden or orchards, parks were to make up for the missing vicinity to nature. In the words of a young mother at Bi'sat Park,

> We usually come to the park because of the kids. Our apartment is tiny and they can't play freely. And we don't have any space for kids in our block. So they are constantly making noise and driving our neighbors nuts. When we come to the park we can spend time together and I can play with my kids. They run around happy and that makes me and them feel good.

The flow of space in Iranian traditional architecture is perceived as an essential component of Iranian culture. The flow takes place through the interconnectedness of various spaces which in turn is expected to ensure a natural movement,

Figure 2.22 Enclosure at the gender-mixed Bi'sat Park in Tehran.

starting from the outer open space of the courtyards toward a semi-closed space of the *dalan* and the closed space of the inner house. Thus courtyards, as the open space within houses, are part of a continuum of various spaces (open, semi-closed and closed) in traditional Iranian architecture. Many family activities take place in such open spaces, which exist as areas in between the public outer space and the private inner. Children play in the courtyards while occupants of the house meet with their networks of neighbors, friends and extended families in small gatherings – particularly in warmer seasons – to spend an evening or so chatting, eating or engaged in similar activities. Western lifestyle, however, introduced small apartments, mostly in high-rises, which deny access to such safe and comfortable spaces as courtyards. In addition to the lack of access to such open spaces in Iranian Western-inspired contemporary houses, the rapid pace of urbanization has resulted in higher population density, crowded streets and long commutes, which in turn have made it harder to reach the countryside and the nature outside of the city. Urban parks have replaced such accessible green spaces which conceptually exist as part of the Iranian lifestyle. Urban parks are among the popular public spaces that provide green open areas for citizens and which attract people longing for closer ties to nature. Hence, the visual attractiveness of parks is among the prime factors attracting visitors to the parks.

Visual attractiveness in the woman-only Bihisht Madaran Park in Tehran

As an old park with dense and diverse greenery, Bihisht Madaran Park is different in terms of visual attractiveness compared to the newly-built Pardis Banvan. Transformed from a neighborhood gender-mixed park to a women-only park, the place's green structure remained intact. The only additions during the transformation process were the construction of the enclosing walls around the park and of some buildings used for recreational purposes.

Generally speaking, most of the park users in Bihisht Madaran Park were satisfied with its green environment. Time and again, they referred to the park's diverse vegetation and admired the beautiful views, which are enhanced by the natural slope and topography in this area. The park's rich green hedges have also been used as an innovative tool to close off the views to the park. In some parts of the park, trees and bushes come together to form a soft and natural barrier separating the space's interior and exterior (see Figure 2.23). Some users bring up the scarcity of convenient furniture and the lack of proper signage and informational boards used to help visitors find their destination as constraints to the park's attractiveness.

Visual attractiveness in the woman-only Pardis Banvan Park in Tehran

As noted earlier, the overall design of Pardis Banvan in Tehran is inspired by the classic *charbagh* design used for traditional Persian gardens. Two perpendicular

Figure 2.23 Entrance to the women-only Bihisht Madaran Park in Tehran.

axes were used to divide the park into four zones, and different activities and functions such as a sport center, a library and auditorium, and an education and health center are placed in each. Nevertheless, in designing the park buildings, modern and traditional architecture principles were combined to achieve maximum style and functionality. A walking path and a biking track, surrounded by trees and plants which belt around and across the park, have created a relaxing and comfortable environment for users.

Variation in furniture, both in terms of form and function, is another interesting feature which has enhanced Pardis Banvan Park's visual quality. There are a significant number of seats of various types, forms and sizes available in the park, which invite visitors to spend time there for various purposes and occasions. This variation can also be observed in the furniture's materials and colors, which lend a more positive visual impression to the visitor and enrich the park's image.

Despite the designers' rather successful efforts in creating an enjoyable, beautiful and relaxing park environment which is noted and appreciated by many, there are some concerns and comments raised by park users that are mostly aimed at improving Pardis Banvan Park. A young woman in Pardis Banvan was particularly concerned about the place's maintenance. Pointing at the lake she says,

> At the beginning, the lake was very well maintained. It was very clean and looked beautiful. But look now. The level of water has decreased and it smells awful.

Talking about the park's image in winter, and reflecting on the fact that the quality of an urban green space should be responsive to different seasons, another woman notes,

> It's summer now, it's warm and nice. Trees, flowers, people have turned this park to an extraordinary place. But just imagine the winter, when there are no leaves left on the trees and no plants or flowers, and not so many people in the park. What an awful place to be in.

Another decisive factor that affects a women-only park's visual attractiveness is the way designers block views to the space and create only a single point of access to the park at the entrance. The entrance of Pardis Banvan Park is creatively designed using several layers of walls to provide gradual (yet concealed) entry into the park (see Figure 2.24). The enclosing walls, however, are made out of simple bricks and are lacking in any sense of creativity, a classic simple solution to enclose any given space without any sense of aesthetic. It is a rendition of *dalan* which connects the public outer space (*biruni*) to inner private (*andaruni*).

Visual attractiveness in the woman-only Pardis Park in Isfahan

Reflecting on the visual attractiveness of the women-only Pardis Park in Isfahan, the users were neither satisfied by its natural greenery nor with its man-made

Figure 2.24 Entrance to the women-only Pardis Banvan Park in Tehran.

structures. The majority of park users describe the park's landscape as "very poor" and one Isfahani woman argues critically, "my private courtyard is much more beautiful than this place and I don't know why they would not even make efforts to improve the quality of this place." Talking to park users of different back-grounds, a clear gap emerges between an ideal image of a park and the actual place at their disposal.

Researchers conducting observations in Pardis Park in Isfahan verify that the park's buildings are not visually attractive. Prefabricated structures have been used to provide spaces for educational purposes and other indoor activities. Walls around the park are composed of three layers placed on top of each other: an old brick wall provides the foundation for large prefabricated asbestos plates with sharp spikes on the top to prevent anyone from entering the park over the walls. These walls are also extremely unattractive, and give the place a fortress-like appearance, which has become the subject of jokes and satire among women (see Figure 2.25).

Visual attractiveness in the woman-only Park Banvan in Rasht

In terms of visual attractiveness, the small and young Park Banvan in Rasht can-not be compared to the other park examples. The park's young trees fail to create an image of an urban green space (see Figure 2.26). Being situated in the midst of the green belt around the Caspian Sea makes it hard to differentiate the visual impression of the park from its surrounding area, particularly given its size and

Figure 2.25 Entrance and visual attractiveness at the women-only Pardis Park in Isfahan.

the lack of vegetation variety within its interior. The park is also lacking any appropriate and visually attractive furniture, signage, landmarks, monuments or any other attribute to improve the place's visual quality. While the city is a joyful place because of its pleasant climate and diverse vegetation, the park fails to impress park users as a green urban space. The dearth of any other leisure activity in the park also contributes greatly to this image.

Visual attractiveness in the gender-mixed Niavaran Park in Tehran

Niavaran Park is one of Tehran's oldest parks and consists of a neighboring cultural and art complex (*Farhangsara-yi Niavaran*) as well as a historical palace and museum (*Kakh-muzi-yi Niavaran*). Proximity to such public functions, each with its own large open green spaces, has bestowed an aesthetic character to Niavaran Park and created a delightful and pleasant green environment. The park's overall plan is inspired by the concept of the Persian garden (*Bagh-i Irani*) and is designed accordingly. Reflecting the image of the Celestial Garden on Earth, all the elements making up the Persian garden are utilized to create a pleasant and relaxing environment. Two principal natural elements of any Persian garden are water and sunlight which, in combination with the diverse greenery of various shades and shapes, form the space's main structure. Those elements and their

Figure 2.26 Entrance and visual attractiveness at the women-only Park Banvan in Rasht.

respective impact on a space have always been taken into consideration in design-ing the gardens inspired by Persian vernacular architecture. The aim is to create a variety of spaces to be used during different parts of the year, not least during the summer.

Moreover, the overall forms of these gardens always follow geometrical prin-ciples, and mostly those set forth for rectangles. Water, both still and running, shapes the space's backbone and divides the space into symmetrically partitioned smaller spaces. Designed and constructed some 50 years ago, Niavaran Park still stands as an exemplary instance of translating the Persian garden's traditional principles into that of a modern public park. The park's main design encompasses two perpendicular axes. The longer axis, which is made from a running water canal, crosses through the length of the space and ends at the park's main body of water. The ground's natural slope and the park's multi-level design make the canal water flow, and the running water's sound, create a pleasant and relaxing ambiance. Additionally, trees and bushes are planted on both sides of the canal at regular intervals and define the walking paths parallel to the park's waterways.

The park's water features, both still (pool) and running (canal), significantly contribute to the space's visual and functional quality. The pool and its fountains at the end of the canal function like a physical landmark, which creates an active node in the park. Unlike other parts of the park, the pool is situated in a vast open and flat area. This quality attracts many park users (particularly the elderly) who enjoy walking and jogging in the area. The diverse vegetation, the variations among furniture, the existence of fitness equipment and proper lighting have turned this part of the park into one of Niavaran Park's most popular spots. Time and again park users indicate that they

> prefer to walk around the pool, since unlike some other parts of the park, it is flat. The running fountains make the air around the pool cooler and more pleasant to enjoy. The sound of the running water also turns walking to a relaxing and joyful exercise.

Avenues of tall trees also mark the park's borders and separate it from the outer urban fabric. Located on a hillside, the park is composed of several smaller spaces built in multi-height levels connected through stairways. The differences in levels and connecting stairways are among Niavaran Park's main visual components. Variations in height, size, form and function of such places within the larger space improve physical diversity and add to the park's visual attractiveness. The park's multi-level design and dense greenery not only lend themselves to forming part of a beautiful landscape, but they also create pleasant and comfortable shady spots for Tehran's hot summer days.

Furniture of various kinds is used in different parts of the park. This diversity in furniture forms, colors and functions has improved the park's quality and facilitated people's use of the space. Benches, chessboard tables, flower boxes and fitness equipment are all painted in vibrant colors and are among the furniture which creates a beautiful harmonious combination with the park's green environment.

Visual attractiveness in the gender-mixed Bi'sat Park in Tehran

While Niavaran Park in northern Tehran is surrounded by various green spaces and is in perfect harmony with the surrounding urban fabric, Bi'sat Park in the south stands in sharp contrast with its neighboring fabric. Despite the fact that Bi'sat Park has lost its glory and importance, as the first public park in southern Tehran, it continues to attract a larger group of users and creates a microclimate to improve the air quality of the neighborhood for many decades to come. The park's overall plan is inspired by the "naturalistic design" principles of the English garden; a landscaping style that emerged in England during the early eighteenth century. As an idealized view of nature, the English garden style fundamentally differs from the geometrical and symmetrical design of the Persian garden.

Bi'sat Park's size has provided a rich diversity of forms and functions in the park's inner spaces and has opened up new possibilities for extending the principles and the elements of the English garden design. To name but some, these include the lake and the bridge over it, vast lawns, groves of trees, recreational spaces and constructions. All these elements are used extensively in Bi'sat Park to create picturesque landscape. As part of a naturalistic landscaping style, curved lines and non-geometrical forms shape the park's spaces, landmarks and elements. Two artificial water bodies designed to look like natural lakes, along with irregularly planted shrubs and trees left to grow to their maximum size and kept in their natural overgrown form, create the park's main visual characteristics and lend themselves to being seen as a series of natural vistas across the park.

The park's soft landscaping coupled with the buildings and the park's outdoor activities have formed an interesting combination of natural and man-made environments. Two artificial lakes providing boating and fishing activities, an open amphitheater, an amusement park, an old airplane,[2] fitness equipment and furniture in various forms and colors are among the main attributes creating the park's overall image. In recent years, the establishment of new parks in the surrounding neighborhoods has affected the number and range of park users. Once popular and crowded, Bi'sat Park is losing its visitors to these newly-built and more appealing parks. The variety of activities and mixed-use within one and the same space is one of Bi'sat Park's main strengths; however, its lack of safety and the insufficient maintenance of existing facilities and equipment is threatening the park's vitality and is resulting in the loss of visitors.

Notes

1 Entering a Persian house, one found oneself in a *dalan*, a passageway which gradually leads a guest into the appropriate part of the building. It also functions as a buffer zone, where the male guest deliberately slows down the pace and chants God's name (*Ya Allah*) to let the female members of the family evacuate the *biruni* or the shared spaces such as courtyards. In more economically moderate families where there was no structural division between *andaruni* and *biruni*, a thick canvas curtain marked the *andaruni* part. The ritual of entering the space, however, was practiced with an equal degree of seriousness. For more details see: Djamalzadeh (1985), Morier (1835) and Benjamin (1887).

2 People tell stories about the airplane in the park which belonged to a member of the Pahlavi royal family. Many years ago, the airplane had to make an emergency landing due to a technical failure. The park was not built back then; but the airplane remained in the same place even after the park's establishment and is used as one of the attractions of the park.

3 Functional dimensions of the women-only parks

with Masoumeh Mirsafa

> Men are conditioned beings because everything they come in contact with turns immediately into a condition of their existence. The world in which the *vita active* spends itself consists of things produced by human activities; but the things that owe their existence exclusively to men nevertheless constantly condition their human makers.
>
> *Hannah Arendt, The Human Condition (1998: 9)*

This chapter discusses the functional dimensions of women-only parks, which involves how parks as urban entities work and in what way urban designers make gender-specific spaces to respond to the needs of the users. Carmona and Tiesdell (2007: 211) argues, "the 'social usage' and 'visual' traditions of urban design thought each to have a 'functionalist' perspective." Departing from this premise, when discussing the functional dimensions of the parks, this two-fold model will be utilized to explain the functionality of the parks. The parks as public urban spaces are created to facilitate social interaction. The social usage is defined and conditioned by the way the designers define and design the parks and the very way that people appropriate and use them. The visual tradition, however, focuses on "the human dimension and how it is often abstracted out and reduced to aesthetic or technical criteria features." As Banerjee and Loukaitou-Sideris (2011: 165–6) put it,

> for urban design, the meanings should be consonant with the functional goals of the place for the public experiencing it. Any design varies in the likelihood that it will evoke a specific meaning among people experiencing it. … Urban designers can use those shared meanings to craft designs compatible with purposes of settings for many users.

A functionally successful place supports and facilitates various activities, and the design of urban spaces should be informed by awareness of how people use them. Accomplished urban designers generally develop a detailed knowledge of urban spaces, places and environments, based upon first-hand experience (Carmona 2003). Whereas, through planning, urban designers define the intended functions of a place, users may modify and appropriate it to their needs and activities. A place, which is designed in agreement with the needs of its users, invites users to engage with the space at different levels. A passive engagement results in a sense of relaxation and enjoyment, while an active engagement involves a more direct experience and interaction with place and people. In a park, for instance, while there is the possibility of active engagement, it may be utilized merely to bestow a sense of relaxation and enjoyment in a form of passive engagement. Hence in Carr's (1992) words "some people find sufficient satisfaction in people-watching, others desire more direct contact." One should note that "the simple proximity of people does not mean spontaneous interaction." Functional dimensions are also conditioned by what Lefebvre and Goonewardena (2008: 137) calls "ideologies of space" which articulate the spatial structures with the practices of everyday life, to render spatial practices coherent, guarantee the functioning of activities and prescribe modes of social life in any given space.

Functional dimensions in this chapter are discussed under:

- "mixed-use," to address how diversity in usage and the possibility of various activities can influence the overall usage of the parks and whether the notion of mixed-use is different in women-only parks compared to gender-mixed ones;
- "adaptability," which explores how (both women-only and gender-mixed parks) adapt to various social and temporal qualities; and
- "management and surveillance," under which the notion of surveillance, control, safety and security of the women-only verses gender-mixed parks are discussed.

Mixed-use in women-only parks

The possibility of a mixed-use for an urban space has always been regarded as a key factor to determine a successful urban space (Jacobs 1969; Montgomery 1998; Hildebrand 1999; Davies 2000; Krier 1990). Mixed-use is a widely acknowledged principle among urban planners, with the premise that the "variety of uses is the key to variety as a whole" (Bentley *et al.* 1985: 27). It is defined as the extent of development on a given piece of land, which in combination with density can improve the vitality and viability of the place (Williams 2000: 43). Enhancing diversity through a combination of different uses is expected to attract different groups of people to a place at different times, and for various reasons. Parks as public spaces are used by a wide array of people from different socio-economic backgrounds who are expecting to find different ranges of activities and functions according to their interest. Montgomery (1998) argues that diversity in activities would ensure presence of people in an urban space across different times of the day, and people would use the place for a variety of different purposes for different reasons and also be able to utilize many facilities in common.

Urban parks are multi-functional complexes, comprised of various types of activities and hence able to attract different groups of people to use the space for a variety of purposes. Well-designed public spaces are those which provide opportunities for people to utilize the place according to their personal preferences and interests. Hence, due consideration should always be directed to create a successful "mix" of different activities and functions within the space to assure variation among users. Such functions should be compatible, complementary and support each other. In this way, different functions will interact positively, complement each other and attract more users of different interests and backgrounds and ultimately contribute to a greater vitality of the place. Proximity to different functions, on the other hand, can provide a greater range of services to respond to diverse needs of users. Table 3.1 shows the type and frequency of activities in parks (without specifying whether mixed or women-only) among a random sample of 1,161 (all age groups) Tehrani inhabitants.

Meanwhile, there are no statistics to demonstrate the trend in the women-only parks; however, these parks offer a range of activities to their target groups, which vary greatly from one park to another. Vast green areas in Bihisht Madaran, for instance, are often used for picnics and group gatherings. Biking and walking tracks, indoor and outdoor sport facilities, children playgrounds and cafés are among other functions at the park. There are also a number of regular programs (recreational, educational and otherwise) and occasional events such as different workshops, educational seminars and seasonal ceremonies, which take place at the park occasionally.

Jacobs (1969: 155–7) argues that "the vitality of a space is maintained through overlapping and interweaving of activities … and its understanding requires dealing with combinations of mixtures of uses as the essential phenomena." The diverse range of activities in the parks is used to create a space to attract a wide range of users of various socio-cultural backgrounds, with different needs and a variety of interests. Functions – including services and facilities – at Pardis Banvan show a greater range of diversity compared to all other women-only parks

Table 3.1 Various activities for which Tehrani respondents use the parks

Activity	Frequency	Percentage
Being outdoors	245	24.2
Sport/jogging	209	20.7
Walking	187	18.5
For kids to play	159	15.7
Enjoying green spaces	69	6.8
Sitting and resting	54	5.3
Picnics and eating	46	4.5
Meeting friends	21	2.1
Others	13	1.3
Reading	9	0.9

Source: Tehran Municipality Office of Social and Cultural Research (2011).

studied. The design of the park is more recent and designers have taken into consideration the needs of women, and endeavors have been made to provide an appropriate space for various activities. Several indoor and outdoor sports facilities (soccer field, and tennis and badminton courts); walking and biking tracks; a botanic garden (specialized in growing flowers and training various skills in gardening and flora artistry); an artificial lake; a cultural center; an educational and family consultation center; a kindergarten and children's playground; a swimming pool; and sunbathing facilities, are among the serving functions at the park. Water-related activities lend this park special features compared to other women-only parks. The lake makes it possible for rowing, while three different swimming pools (two for adults and one for children) provide a wider range of possibilities for the visitors. Reasonable ticket prices compared to similar facilities in the city along with its rooftop sunbathing facilities have turned the swimming pools into one of the most popular services in the park and they attract many users.

The variety of uses (mixed-use) is strongly influenced by the size of the space. As the space shrinks, the diversity of its facilities diminishes accordingly. Pardis Park in Isfahan is much smaller in size compared to the two parks in Tehran and has a limited possibility of providing services to its users. Besides a green open space, Pardis Park has outdoor and indoor sport facilities, a biking track and occasional market for homegrown products and artifacts. The park is also used for various ceremonies and gatherings of different kinds.

Due to its small size and limited functional capacity Park Banvan in Rasht, fits into a different category. It could safely be called a "pocket park," which is distinguished by its small size, yet accessible to all. Park Banvan is well interwoven into the urban fabric in the most developed part of the city. As with any other pocket park, it is too small for physical activities and functions as a small microclimate in the midst of a well-developed urban fabric. It is composed of a green space and users can define and decide the usage. It is used as an outdoor sitting place and, upon the presence of children, as a playground. There is also a medium-size sport hall in the park, which was closed during multiple observation sessions and hence it was hard to determine what kind of activities and what occasions the hall is used for. Difference in the mixed-use of the parks demonstrates that for a space to be vital and attractive, it "must serve more than one primary function ... and there must be a sufficiently dense concentration of people, for whatever purposes that may be there" (Jacobs 1969: 162).

In general, users of the women-only parks in Tehran are satisfied with the overall functionality of the parks; however, they do argue for improvements to be made. In all parks women complain about the lack or insufficient number of facilities and services such as restaurants, cafés, food venders, toilets, drinking-water fountains, furniture and picnic facilities. On many occasions users are critical about the misplacement of the facilities (discussed partially under physical dimensions). Women in the Pardis Park in Isfahan demonstrate a lower degree of satisfaction about the parks and their amenities. They are, however, more concerned about the existence of a park devoted to women, a space which belongs to them and they "are not questioned by the male members of the family" for being in the park.

However, women are also aware of their needs and demand them. They critically reflect on the lack of facilities and possibilities for activities in women-only parks. Many women argue that, most of the facilities in these parks are also available outside. However, they prefer to use those in the parks while enjoying the aesthetics of the park and experiencing a sense of comfort and developing a sense of belonging to the space. In-park facilities in women-only parks also make it possible for women to meet in groups and spend time together. The sense of comfort and feeling of freedom along with a wide range of amenities turns the parks into a pleasant place to visit and the experience promotes a wish to return to the place again. A young woman in Bihisht Madaran notes, "in this country, men feel comfortable in all parks but it is only in these [women-only] parks that women feel comfortable." Another woman talking about the benefits of such a park notes, "It's really good to have such places. After all, our lives are shaped by religious values. We don't like to expose our bodies to men, you know. But we need a place to set our energy free."

To maximize the usage, women-only parks need to attract different groups of women to their various activities and services at different times. Services are expected to respond to the needs of all women with different economic, social and cultural backgrounds. Designing a space for women requires attention to the variation and needs of different groups of women. While diverse in providing activities and facilities, women using the park note a visible flaw, namely the needs of young children who accompany their mothers to the parks. Numerous references by mothers were made to the lack of safety for younger children. A young mother at Bihisht Madaran notes,

> This park is not child-friendly at all. There are not enough amenities and services for kids – just two small playgrounds with some simple toys which lack safety standards. You get the sense that mothers with children are completely ignored and not welcomed.

Promoting diversity through a combination of different functions is likely to attract various groups of users and improve the vitality of a public space. Variety and a mix of uses at both micro (inside the park) and macro scales (surrounding areas) are among the decisive factors in creating a responsive urban space. While women-only parks in many cases suffer from incompatibility with the surrounding fabric, both gender-mixed Niavaran and Bi'sat Parks are located in mixed districts and surrounded by various urban functions. Niavaran Park is located in an urban district with a higher cultural and socio-economic status, predominantly residential and related services. Residential blocks surrounding the western side of Niavaran Park and alleys in between provide easy access for the locals to reach the park. The historical palace and museum (*Kakh Muzeh-yi Niavaran*) in the north, the cultural center (*Farhangsara-yi Niavaran*) in the east and Tehran Research Center (*Markaz-i Mutalia'at-i Tehran*) and Shahr-i Kitab bookstore on the southern side, in combination with the park itself, have created a socio-cultural complex in northern Tehran that lends a strong cultural identity to the neighborhood. Moreover, compatible functions and related activities have enhanced the use of park and contributed positively to an improvement in the quality of public life in the area.

Although providing mixed-uses is one of the essential requirements of a well-designed and functioning public space, due consideration should be given to providing a compatible and complementary "mix" of different activities and functions in the space. The neighboring functions in a well-mixed district should be compatible in ways that support other activities (see Figures 3.1, 3.2 and 3.3, cf. Figures 3.4 and 3.5). Bi'sat Park in the south of Tehran, for instance, is located in a district with a mixed urban fabric. Offices and administrative bureaus, a grand bus terminal and small industries surround the park. Presence of such functions surrounding the park has affected the overall quality of the neighborhood and has diminished the attractiveness of the park for users. The type of the functions and activities neighboring Bi'sat Park have lent a different character to the park and surrounding neighborhood compared to that of Niavaran's. Whereas Bi'sat Park is located in an area surrounded with industrial and incompatible functions, the residential facilities and related services as well as amenities of city-wide significance (cultural and otherwise) in Niavaran Park have transformed it into an institution with a proper mixed-use. Such a complementary combination of uses would form a desirable image of the park, improve the perception of the place and encourage people to use the park (see Figure 3.5, cf. Figure 3.6).

A mix of uses in micro scale (inside the park) has a major role in responding to the specific needs of users. By providing possibilities for doing various activities inside the park, users enjoy spending time there. Having pleasant and joyful experiences creates good memories, forms images of the place in the minds of the users and encourages them to return and visit the park. Gender-mixed Niavaran Park is a good example of such a place. Different types of users (in terms of age, sex, needs and interests) seek satisfaction with the functions and facilities of their interest provided in the park. In addition to the large green- areas, a foreign-language education center (*Kanun-i Zaban*) and a multi-purpose playground provide space for activities for children and teenagers, along with a café, and indoor and outdoor sports facilities to enhance the mix of uses on an in-park scale. Given the size and multiple functions of Niavaran Park, such a mix of uses and functions inside the space has resulted in a greater diversity among users, and has bestowed more vitality to various parts of the park at different times of the day.

Bi'sat Park also provides a wide range of functions and activities inside the park. The amusement park (which was closed during the observation sessions); children's playground; indoor and outdoor sports facilities (a soccer field, and basketball, volleyball and badminton courts); a branch of the popular Institute for the Intellectual Development of Children and Young Adults (IIDCYA);[1] a mosque; and food corners are among the places and activities inside the Bi'sat Park. Despite the fact that the large size of the park and variety of activities and uses inside the park might be responsive to the needs of a wide range of users, Bi'sat Park is not considered a popular public park among users. Due to the age of the park and insufficient maintenance, some facilities are broken or out of order

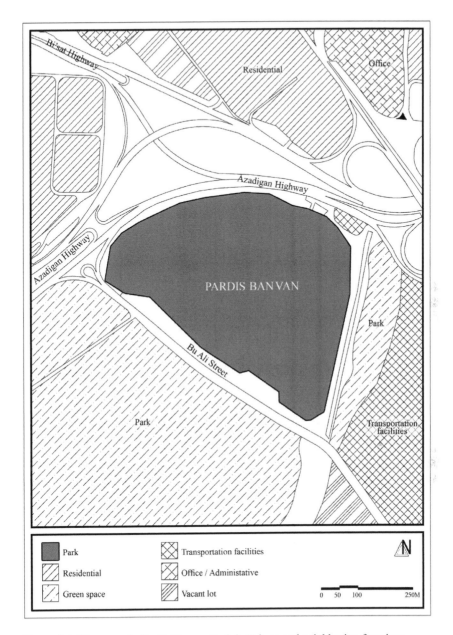

Figure 3.1 Women-only Pardis Banvan Park in Tehran and neighboring functions.

while some other lack the required safety standards. As noted earlier, the amuse-ment park had stopped working a few years earlier. Despite the space devoted to run such activities and the available infrastructure inside the park, an absence of renovation and replacement arrangement has resulted in dysfunctional services.

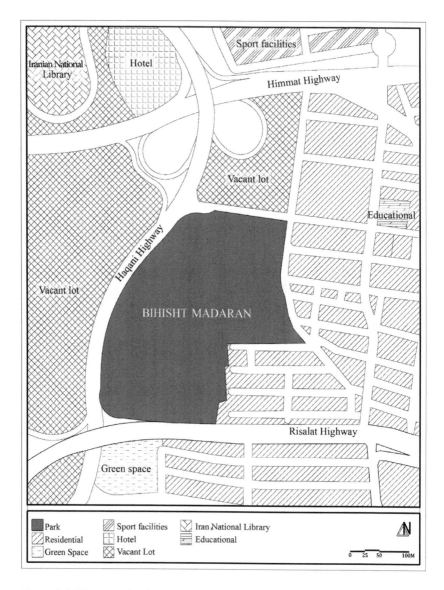

Figure 3.2 Women-only Bihisht Madaran Park in Tehran and neighboring functions.

For instance, an artificial lake, one of the main attractions of the park, popular for its rowing and sport fishing, is not in use any longer. Users repeatedly refer to it as an example of mismanagement and lack of maintenance which results in loosing park attractions. When interviewed, many note that the park was functioning better in the past and it is in desperate need of improvements to extend services

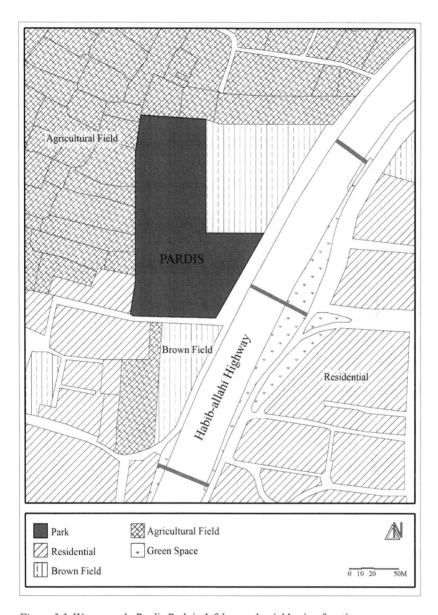

Figure 3.3 Women-only Pardis Park in Isfahan and neighboring functions.

in order to attract users. Such instances and many similar examples, which are keenly noted by park users, are among the main reasons for the lack of users' presence in the park, which significantly affects the sense of vitality and liveliness of the park.

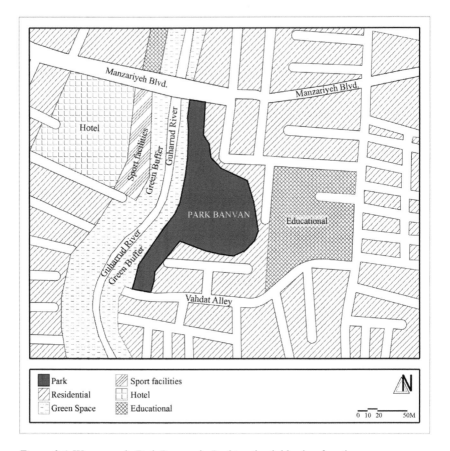

Figure 3.4 Women-only Park Banvan in Rasht and neighboring functions.

Urban parks provide an appropriate space for different groups of people to define and use the space in their own way. This feature, also known as adaptability, offers different choices to users and provides the possibility for extended and innovative ways to use the space. While this may happen at an individual level, it can also be practiced on a public scale. One of the most interesting examples of adaptability in urban parks is using the parks as a space for holding public events. Seasonal festivals, traditional and religious celebrations are among the public events that usually take place in the parks several times a year. Such events are usually (semi-)official and often held by a municipality or other similar organization.

The vitality of a place is also attributed to its adaptability, which is a quality that allows a space to change easily and hence remain sustainable. It is a "capacity of a building or space to be changed so as to respond to changing social, technological and economic conditions" (Carter *et al.* 2015: 41). While the term adaptability is widely used for such a quality, some scholars use the term "robustness"

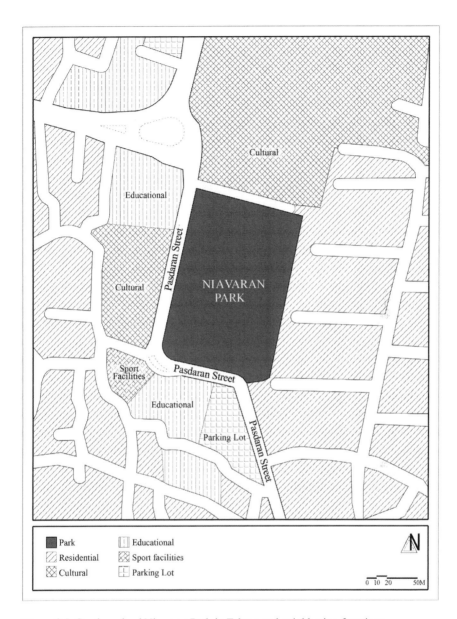

Figure 3.5 Gender-mixed Niavaran Park in Tehran and neighboring functions.

interchangeably to explain a quality which refers to an attribute that make "places which can be used for many different purposes, offer their users more choice than places whose design limits them to a single fixed use" (Bentley *et al.* 1985: 10). The degree of adaptability and robustness is conditioned by the form and functions of the spaces. The notion of adaptability also connotes that through

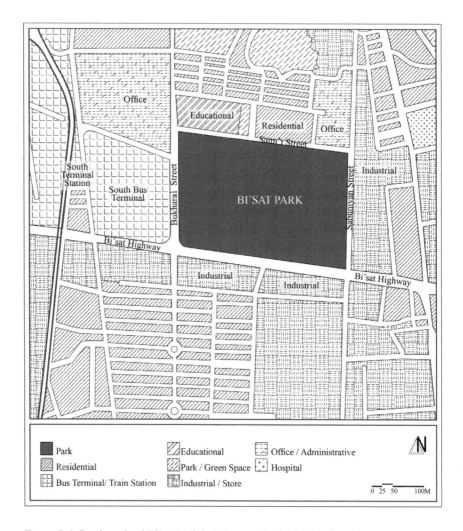

Figure 3.6 Gender-mixed Bi'sat Park in Tehran and neighboring functions.

moving the border between inner parts of space (sometimes indoor) and the outer space (outdoor), one can extend the inner space and the activities within it. Hence, "activities within the space may benefit from being able to extend outwards into adjacent public outdoor space. When this occurs, they will contribute to the activity in the public space itself" (McGlynn *et al.* 2013: 58).

Adaptability at women-only parks provides the possibility of extra uses of the parks for special occasions or events. In other words, it allows women to use the place in their own way. This adaptability, on collective/personal, official/non-official levels, and both on macro and micro scales, invites and encourages women to use the park. Findings of a study (Tehran Municipality Office of

Social and Cultural Research 2011) reveal that some 80 percent of the population of a representative sample from Tehran perceive "proximity of a park to cultural and educational centers" as a "very important or important" criterion in choosing a park.

Holding seasonal festivals, temporary exhibitions, markets, traditional and religious celebrations in Pardis Banvan, Bihisht Madaran in Tehran and Pardis Park in Isfahan are examples of such adaptability of place, which in turn improves the vitality of the place and introduces the park to new groups of women.

Adaptability may also be used at a personal level and in combination with personalization of the space. An individual or a group of users may decide to hold informal weekly meeting on the lawn at the specific space. Though the lawn by its function is not defined as a meeting place *per se*, it is adapted to a new usage. Examples of such use are further explored under social dimensions and network building in Chapter 4.

Surveillance and management

The notion of surveillance and control is among the primary concerns of any public space and is regarded to be as crucial as the exercise of power and provision of safety and security. The women-only parks, as ideology-laden spaces, are subject to, among other things, "the organized practices through which subjects are governed," and also to the Foucaultian practice of biopolitics and governmentality. Hence, public spaces become like "enormous Panapticons" (Tabor 2001; Fyfe and Bannister 1998) in which "an inspecting gaze, a gaze which each individual under its weight will end by interiorizing to the point that he is his own overseer, each individual thus exercising this surveillance over, and against, himself" (Foucault 1980: 155). While the notion of surveillance could be explained in terms of panoptic control and hence practice of power, it is a serious concern to be addressed by urban planners in designing any given public space. The delicate task of creating a balance between the civil rights of the citizens and the surveillance of a space is easier said than done. In Carmona's (2003: 125) words, "the public realm often needs to be managed to balance collective and individual interests. This inevitably involves finding a balance between freedom and control. Freedom of action in public space is, nevertheless, necessarily a 'responsible freedom.'" This proposition, however, suggests that the extent of freedom and control varies from one place to another and is conditioned by the characteristics, functions and regulations of any given space. A rather commonly consented taxonomy of control among the scholars of urban studies is suggested by Loukaitou-Sideris and Banerjee (1998: 183–5) who classify control into:

• hard (active) control, which utilizes security officers, surveillance cameras and regulations either to prohibit certain activities or to allow them subject to the issue of permissions, programming, scheduling or leasing; and

- soft (passive) control, which focuses on the "symbolic restrictions," and passively discouraging undesirable activities or on refusing to provide certain facilities.

Management of the women-only park has invested in the means of both hard and soft controls. A sophisticated system of "hard" control is installed to ensure that the normative principles outlined by the religious regime for women-only spaces are achieved and only the permissible activities and conducts in the public domain are practiced. A combination of various controlling strategies, including security officers (both at the entrance and those patrolling inside the parks), a network of surveillance cameras, and a series of regulations which prohibit certain activities and conducts within the space, is pedantically implemented. While some strategies are consistently implemented and practiced in all parks, the level of hard control and strategies for soft surveillance vary from one park to another.

Certain regulations give the impression that the main concern of the authorities is the control of the visitors and hence the practice of power, rather than any rationale – whether instrumental (*zweckrational*) or value-/belief-oriented (*wertrational*), to use Weberian categorization. For instance, carrying cameras or any device which can take photos (such as mobile phones) is strictly prohibited in the women-only parks. However, in Tehran, a professional photographer is in place to take photos of the visitors on exchange for a payment. That said, some 88 percent of women visiting women-only parks were concerned that they might be filmed or their photos might be taken by fellow women inside the parks. Interestingly, the next biggest concern was the presence of transsexuals or cross-dressed men in the parks (Kawsari 2008: 114). Women are asked to hand in their forbidden items at the entrance where a number of deposit boxes are installed. The checking process upon arrival at the parks also varies from one park to next. In Bihisht Madaran the security officers at the entrance ask women to observe regulations of the park and hand in the prohibited items voluntarily, while in Pardis Banvan all women queue waiting for inspection to be administered individually and strictly by the officers, both bodily and throughout their belongings. In Isfahan's Pardis Park the process is somehow similar to that of Tehran. In Park Banvan in Rasht there is no checkpoint at the entrance since the park is not enclosed and women can freely enter the park without any search. There is only a small booth at the entrance occupied sometimes by a male security guard. There are two signs, each installed at either side of the entrance. The text on one sign indicates, "Women-only park. No entry for men," and on the other side another sign of a similar size and font indicates, "Animals are not permitted in the park." The combination of two signs has been a subject of jokes by park users and passersby.

While gates equipped with surveillance systems and an elaborated checking process may enhance a sense of security among some users, others perceive it as an indication of distrust and disrespect. A woman in Pardis Banvan who demonstrates no concern about going through the inspection process, says, "I think it's a necessary procedure and I respect it. It's for the best of everyone to have a safe and secure park." Another woman in Pardis Park in Isfahan, however, considers it disrespectful and holds, "they check us so carefully at the entrance that makes me feel that I'm about to enter a military zone, not a park." Some women do not show any objection to the security check *per se*, but to the way they are administrated by the officers. A young woman in Pardis Banvan reflects,

> I know that it's part of the regulations and they have to check everybody to make sure that everything is ok, but they can do it in a much nicer way. I don't like the narrow pathway at the very entrance to keep women in a single line. It's so annoying and disorderly …
>
> I think they need to devote more space for the checkpoint or introduce at least two or three spots for such occasions when a few women arrive at once, so they don't have to wait in a long line to be checked.

Male children older than five years are not allowed to company their mothers into the parks. Many younger women find it a serious restriction, which prevents them using the park. Smoking is strictly forbidden inside the park and sunbathing is only possible at designated places. The park is under the constant "inspecting gaze" (Foucault 1980: 155) of patrolling security officers and surveillance cameras at all times.

Hard controls tend to be top-down, formal and law-oriented; soft controls are informal, subjective, spontaneous and peer-to-peer – inspired by and practiced under the Islamic tenet of *amr-i bi ma'ruf ve nahy-i az munkar* (commanding right and forbidding wrong).[2] Soft controls are formed around the religious values and cultural norms, and are in line with the objectives defined for a public space in a Muslim society in harmony with collective morals and shared values of the community. It could also be informed by *'urf*. The patrolling moral police and the fellow women are not the only ones to advise visitors to observe their religious and moral duties, a series of signage, murals and billboards inside the park constantly remind them of the normative and religious values of society. The messages usually advise women to abide by the behavior expected of women in an Islamic society. One billboard inside Pardis Banvan reads,

> Respected ladies, since some areas of the parks might be visible to the neighboring buildings overlooking the park, please do avoid revealing dresses and observe a modest dress code while in the park.

And signage in Bihisht Madaran states,

> Even though the park is a women-only place, please do respect the social and religious values by not wearing too short, too tight or too revealing clothes.

The patrolling guards in the parks are responsible for watching over the park visitors and making sure that regulations are observed. They move around the park and remind visitors of their moral obligations on regular basis. They engage in discussion and call the police if they see inappropriate behavior or if they are ignored or meet with resistance from the visitors. Many women in the parks, being used to such encounters everywhere in public, seem not to be bothered by the guards' attitudes. When asked, they usually think they have been treated respectfully. "They are here to remind one of her religious duties and demand her to show courtesy to others. Doesn't bother me much," says a middle-aged park visitor.

Fashioning a balance between the civil rights of an individual in a public space and surveillance of the space is among the key factors in creating a successful and responsive urban space. However, an individual's civil rights and freedom are to a large extent conditioned by the functional structure and guidelines of such place. While the notion of surveillance (both hard and soft) in women-only parks aims to monitor the normative conduct of women within the space and men in its surroundings, the surveillance in the gender-mixed (Niavaran and Bi'sat) parks is ordained to improve the security of the space. Like the women-only parks discussed earlier, Niavaran and Bi'sat Parks apply various methods of surveillance, both hard and soft. The extent and the systems of surveillance around each of those parks are affected by variation in the socio-economic characteristics of the urban surroundings, the history of the place and its image among the public, which in turn affects the pattern of utilization among its users. The Niavaran neighborhood, including the park within it, is usually perceived as a safe public place, while Bi'sat Park has earned a reputation as an unsafe space with a higher rate of criminality. Such a reputation and image, even though unsubstantiated by those who mention them, affect the pattern of use and vitality of such an urban space and result in a declining flow of users into the park. The authorities provided measures to improve the image of the park through investment in the safety issues and removing security concerns. A combination of hard surveillance methods, such as surveillance cameras and a constant visible presence of the police and guards along with a series of regulations made visible through signage and influenced by the police and guards, are meant to bestow a sense of security on the park users.

A local police station is erected at the northern edge of Bi'sat Park to emphasize the presence of law enforcement and provide round-the-clock police surveillance in the park. The strolling police patrols in the park ensure that the regulations will be in place and request users to report any unlawful or inappropriate activities. Such surveillance methods, however, have caused various reactions among different groups of users, especially women, at the park. Users are generally satisfied with the constant presence of the police at the gender-mixed parks and found it an effective way to improve security. Others, however, perceive it as restrictive and in some occasions even disrespectful. Younger users (18 to 28 years old) are among those who consider the presence and occasional inspection of the police at the park a serious obstacle in using the space freely. In the absence of appropriate places and socio-religious restrictions for young girls and boys to meet up and socialize, urban parks are turned into popular spaces. The free access and the possibility of staying longer turns urban parks into one of the most popular alternatives for such group of users with limited resources and possibilities. Compared to other public places such as cafés or restaurants, accessibility at no cost is an advantage of public parks for many younger users especially those with a lower socio-economic background. Younger generations seem to be more aware of and disturbed by the surveillance methods at the park. Some (from all age groups) found the new directives, regulations and surveillance restrictive and in some cases out of proportion. Many women expressed their concerns over those regulations and surveillance methods and express that they are offended by the "distrust" and by being "constantly watched" or "questioned" by police about their presence in the park. A middle-aged woman at Bi'sat Park explains,

> the police are everywhere in the park. I don't like it. I usually come to the park, sit in the same corner and read my magazine. They come and ask me silly questions repeatedly. They are accusing me for no reason. This is ridiculous.

Contrary to Bi'sat Park, Niavaran Park does not use such hard surveillance methods to bring security to the space and to its users. In the absence of an electronic surveillance system and security guards, users, especially those from younger generations, enjoy more freedom in the park. They are not worried about being watched or reminded of their inappropriate behavior or activities. This difference contributed significantly to a contrast between the image of Niavaran and Bi'sat Parks. Many users have noticed that and referred to the lack of police and security guards and absence of surveillance system as one of the reasons that makes Niavaran popular and attractive to younger users. Youngsters (both female and male) from various parts of Tehran commute longer to Niavaran Park and consider it to be a different place compared to many other similar parks across the city and in their neighborhood. In addition to the lack of hard surveillance and control systems, the combination of users from higher socio-economic background and normatively relaxed residents in the neighborhood has made the park an attractive spot in the city. A young man sitting on a bench next to his girlfriend at Niavaran Park says,

> This park is far better than parks in southern Tehran, where I live. People are nice and treat you respectfully. They pass by us sitting here and don't look surprised. But it's completely different in my neighborhood. When you are with a girl, they stare at you in a strange way. It seems as though they are witnessing something weird.

While the new regulations and surveillance systems at the park have raised complaints and concerns among certain groups of users, others are satisfied with them and find it necessary to provide security in the park. Many users note that the new surveillance method has enhanced the quality of the park and improved the reputation of the park as a safe public space. A young woman at Bi'sat Park says,

> Before the police were stationed in the park, it was not safe in here. Many drug addicts were hanging around at the park and this made the park unsafe. At some point we decided not to come here at all. This is the first time we

Table 3.2: Women-only parks – operating hours

	Bihisht Madaran (Tehran)	Pardis Banvan (Tehran)	Pardis Park (Isfahan)	Park Banvan (Rasht)
Operating hours	08:00–19:30	08:00–19:30	08:30–17:00	24 hours
Non-operating hours	Fridays (gender-mixed)	Saturdays closed	Fridays & Saturdays closed	–

have visited the park in a long time. Now I can see a lot of improvements, and I guess it's because of the police presence in the park.

Operating hours of the parks is a serious concern raised by many users of women-only parks. Operating hours of women-only parks are significantly shorter than gender-mixed parks. While many women only have the possibility of or preference for using parks later in the afternoon or in the evenings, the parks are then closed for watering the vegetation and lawns and performing other maintenance, mostly done by men. The traditional division of labor, where men are responsible for manual labor, compels authorities to close the parks earlier so that the male workers can assume their responsibilities. Shorter working hours not only affect the functionality of the parks, but also exclude workingwomen who only have the possibility of attending parks after their daily work. Hence, the parks become unusable for workingwomen, and serve mostly housewives and/or women who come as a group through their school or workplace. Table 3.2 shows the women-only parks' operating hours against the round-the-clock operating hours of the gender-mixed parks. It further notes that Bihisht Madaran is open to public (both for women and men) over the weekends and official holidays and Pardis Banvan is open on Fridays, to provide a slim possibility for workingwomen or those women who are not able to use the park on weekdays for various reasons. Pardis Park in Isfahan has the shortest operating hours (it closes at 17.00) and remains closed on Fridays and Saturdays. The geographical location and the arid climate of Tehran and Isfahan, make late evenings the most desirable time for people to enjoy outdoors activities and to use the parks, when they are closed.

Notes

1 Founded by then the queen of Iran, Farah Pahlavi, in 1961, the Institute for the Intellectual Development of Children and Young Adults (IIDCYA) or *Kanun-i Parvaresh-i Fekri-ye Koodakan va Nojavanan*, a.k.a. *Kanun* has functioned as one of the main centers for the promotion and production of children's literature and art in Iran. With more than 850 children's libraries and cultural centers across the country, *Kanun* is the most active and well-developed network for cultural and artistic activities of children and young adults in Iran. Many famous artists and cultural figures of contemporary Iran started their activities in *Kanun*. *Kanun* libraries are very popular among the Iranian children and youth. Mostly built in public parks to be accessible to all, *Kanun* functions as a magnet to attract children and young adults.

2 The term is used to refer to the exercise of legitimate authority, either by holders of public office or by individual Muslims who are legally competent (*mukallaf*) with the purpose of encouraging or enforcing adherence to the requirements of the *sharia* (Cook 2012). Although the scope and the manner of its practice are extensively debated, it is common across the Muslim world and serves as the theoretical doctrine behind the so-called moral Police in many Muslim countries. For some, the ambiguity in the scope of the term has paved the way for the exploitation and misuse of it in many instances.

4 Women in public parks

Social dimensions

At 13.30,
Me, ... you,
and ... God.

Settled freely,
in Bihisht Madaran.

The sign: "No entry for men,"
not in order anymore.

What a pleasant day it will be.
Sarah Aramesh (2010)

Urban parks are regarded as major contributors to the physical and aesthetic quality of urban life and as venues to facilitate and promote human interaction. Under social dimensions, this chapter endeavors to explore the relationship between people (society) and public spaces, and considers the way people perceive, use and respond to their surroundings as part of their interaction with space. This also means that there exists a "social construction of nature versus the material nature of the environment because the term allows for both. The world is out there, and we interact with it in ways that reproduce it, often altering it in the process" (Smith 2010), yet the world only has meaning for us as language-using and symbol-making animals owing to how we intellectually apprehend it (West *et al.* 2006: 252). Spaces, hence, are not solely physical entities but vessels of human interactions and social activities. Spaces and people are two interrelated concepts and it is difficult to think of space without its social content and of society without its spatial component.

Social dimensions of the space could also be defined as the way

> biologically individuated bodies are situated in locus where they occupy a place. The locus, *topos*, can be defined first in absolute terms as the site where an agent or a thing is situated ... it can also be defined relationally, as a *position*.
> (Bourdieu 1996: 11)

Hence the position comes as the interaction between the space and its consumers (de Certeau 1984). Such interactions, which are the overall idea of this chapter, suggest the process of re-appropriation of space through various *tactics* to adapt the space to one's need.

Informed by methodological approaches in urban studies, social dimensions in this chapter follow a four-fold analysis of the parks as public spaces. "a place for all" addresses the inclusive and egalitarian notion of public spaces and analyzes the parks in terms of their accessibility to all. It also discusses how users of a park as a public space develop a sense of belonging in the process of using the parks. "Vitality" is a key criterion for a public space and argues the manner in which various activities bestow life to a given space and attract new visitors. "Safety" and "security" are probably the most important factors affecting the presence and pattern of usage among women in any public space. Creating a sense of safety within a park – as any public space – is likely to provide a sense of attachment to a place. This is discussed under "sense of place," where methods of re-appropriation of the parks by various groups of women and the notion of belonging are also addressed and discussed in detail.

Women-only parks: a place for all?

Urban planners' strong interest in outdoor urban spaces is – in part – informed by a widely acknowledged fact that such spaces significantly enhance the quality of urban life, promote social inclusion and help to create coherent and functioning communities. This also suggests that public spaces are constituted of two intertwined dimensions: the "physical" (space) and the "social" (interactions). Creation of the physical space, including the material space and settings (both publicly and privately owned), is the main focus of urban planning aiming to facilitate and improve public life and social interactions. Hence the activities and events in the physical space fall under the domain of what Oc and Tiesdell (1997: 18) label the "sociocultural public realm." Publicness of such places indicates that they should generally be at everyone's disposal free of any charge and without any pre-conditions. By definition, the public realm and its institutions, spaces, services and facilities are expected to be inclusive and accessible to everyone. However, some public places, such as women-only parks, are defined to be used exclusively by specific groups of people while intentionally excluding others.

Paradoxically, social inclusion is used as the main argument for creating women-only parks but the creation of women-only parks rests its foundation on excluding men to introduce a new type of public urban space with intended social exclusion. The criterion for inclusion in such spaces is constructed around gender and conditioned by Islamic normative values. Hence, women-only parks function as gated spaces with comprehensive and fully enhanced multi-fold surveillance arrangements. Following an urban fortress model,[1] the flow of visitors into the parks is controlled and filtered strictly. Women are only permitted to enter the park through the main (and in most of the cases the only) entrance that is equipped with layers of security. The entrance area is controlled by electronic surveillance systems, while guards search bodies and inspect the belongings of every single visitor before letting them enter the park. Given that the main objective in designing and erecting any urban public space is to improve the quality of public life through facilitating social interactions among various groups of population, how can one explain the inclusion/exclusion in women-only parks? And, in what ways can the creation of a public space based on social exclusion guarantee inclusion?

The contested notion of deliberate segregation is said to provide women with the experience of a social life in an open public urban space. Islamic codes of social interaction restrict the presence of women in public; and women's presence in public spaces is solely possible by observing *hijab*, through covering hair and veiling the body and limiting interactions to women. In such a situation, however, excluding men from entering the parks provides women with the freedom to remove their *hijab* and to interact freely among themselves. Thus, while on the one hand women-only parks are excluding the male population of the community from accessing and using such spaces, on the other hand this offers women a new opportunity to experience a social life in an open urban space, in the absence any other decent alternative.

A public space for all should not only be accessible to all but should also be available free of charge. Of the four women-only parks observed for this study, Pardis Park in Isfahan is the only park that charges a small entrance fee. This, however minor fee, along with issues in accessibility, visibility and legibility has rendered the park unpopular with women. Many users complain about the entrance fee, especially when they do not receive any services for their payment.

The popularity of urban parks in Iran partially stems from the possibility of access to a wide range of amenities and services, and from being a venue for holding social events and activities. In such a dense and populated city as Tehran, in the absence of private open spaces in houses, lack of time as a result of long working and commuting hours, and air pollution, such public spaces turn into intriguing spots of high quality urban life. Free access to the parks is a source of attraction for a wide range of users (from various backgrounds) and the purpose and frequency of the visits varies significantly from one park to another and from one group to next. Amenities and services at the parks; a sense of safety, security and comfort within the space; good connectivity; easy access; and compatibility of the social fabric of the surrounding urban areas are among the factors affecting the use of space by various groups of users. A successful public space is one that identifies the needs of its target group and satisfies their demands.

Patterns of usage vary among different groups of women in these parks. At both Niavaran and Bi'sat gender-mixed parks, the elderly – mostly men – are among the prime users of the parks. The design of the parks seems to recognize and address the needs of users with limited physical abilities. The practical outcome of such recognition is that the parks become user-friendly not only for the elderly but also for children, families, youngsters and people with physical disabilities. With limited resources devoted to support the elderly and senior citizens in Iran, this group of users tends to take an interest in using parks for their meeting places and socialization. Such needs turn parks into venues for engagement in various kinds of activities and networking. In other words, the publicness of the space provides possibilities for social interaction and lends a sense of inclusion to the community, despite the lack of formal social institutions. The design of the physical space and functional qualities of the space encourages the presence of different groups in public and facilitates intergenerational interactions and social networks in the parks.

Hence, it is not surprising that the parks are among the most popular urban spaces for the elderly in Tehran. Despite the fact that the elderly as a group in Tehran tends to be treated as a homogenous group, one should be aware of in-group variations which in turn result in difference in the usage of public spaces. A significant difference is also observed between women and men in terms of the time, frequency and purpose of the usage. While men are more likely to spend their time in parks, sitting and chatting, women usually visit the parks less frequently and mostly for specific purposes including walking, exercising, spending time with friends or families, and the like. Women, also, usually use the park at specific times of the day. Observations suggest that women are prone to use the parks in the morning and afternoons but seldom in the evenings. Many women note that they feel unsafe in the park after the dark and prefer to visit the park later in the evening only with their family or in the company of a male relative. Such concerns and limitations are not expressed by men.

Contrary to men who use the parks to make new contacts, create their own social network and/or maintain or expand pre-existing ones, women do not usually use the parks to establish contacts or expand their social interactions. They visit the parks with their relatives, neighbors, friends or family members. In response to the question on what hinders women visiting the parks more often, many – including those at the age of retirement – pointed out their responsibilities at home as the main obstacle to using the parks more frequently. Women also repeatedly refer to the traditional normative structure of the society (*'urf*) where men as bread-winners are expected to work outside the home and women remain at home as housewives. By default, even after retirement, men prefer to spend some hours of the day outside, while women stay at home. The never-ending nature of work at home as a housewife, hence, does not provide ample opportunity for women to enjoy their time outside home. Work ceases to exist for men after retirement but not for women, and this seems to be one of the limiting factors for women using the parks.

Young people are also among the main groups to use the park in different ways and for a variety of purposes. With greater flexibility and creativity in using the space, they usually re-appropriate and redefine functions of the spaces and find their own ways to adapt the existing amenities to satisfy their needs. Spending time with friends, dating (different in gender-mixed parks compared to women-only), reading and exercising are instances of various activities that younger women prefer to enjoy at the parks. Such diversity can not only be observed in the type of activities but also in using various parts and spaces within the parks.

In gender-mixed parks, sitting casually on the stairways and edges of the shorter walls surrounding the parks and gathering in the darker and more secluded corners are instances of spaces used by younger male visitors. Such spaces are not attractive to other groups of users like the elderly, families and children. Depending on the purpose, they visit the park at different times of the day. While many male youths use the park to exercise in the morning, some others prefer the pleasant shady and peaceful green milieu for reading or hanging out later in the afternoon. In the mixed parks, more male youths are prone to use the parks for such activities as reading and exercise – both alone and in a group – than females. In women-only parks, however, the presence of young women shows similarities to that of male youths in the mixed parks. They use the park for a series of activities at different times of the day. One may note that legal restrictions along with normative and traditional values which hinder the presence of women in mixed parks do not exist to the same extent in the women-only parks. Young women in the women-only parks experience a sense of freedom similar to that experienced by young men everywhere in a more traditional patriarchal structured society.

Dating is another activity that youths use the parks for. Both Niavaran and Bi'sat Parks are among popular places for young women and men – despite being unlawful and restricted – to meet and spend time together. The number of such users at the parks usually increases in the afternoon and early evening. Young couples who use the space for dating usually avoid crowded and busy parts of the park and prefer the somehow quiet and comfy corners, which are usually less visible and so provide more privacy. Dating has turned into a visible social problem and conservative factions within the Iranian power regime perceive the part of a Western "cultural assault."[2] Such groups demand immediate engagement of the legislative bodies, government and police to enforce law and curb the spread of such

> aberrant moral and cultural problems, and to remedy this serious and infectious disease. This is a termite which has started to consume the foundation of our society … and eats everything from inside and keeps nothing. Not even a mere façade.
>
> (Jomhoori-ye Islami 2010)

Heterosexuality is defined as a norm; and heterosexual relations through marriage as the only permissible form of relationship is institutionalized and practiced. Every random couple is subject to screening and required to present documented proof of the marriage at any point, anywhere. This makes heterosexual dating visible and provokes debates and disputes. However, homosexual dating in the women-only parks does not wake any suspicion. Women-only parks are used as hassle free dating spots by homosexuals. Holding hands, embracing and sitting intimately close to each other in public, which are all forbidden practices between a man and a women without marital connections, are ignored or perceived innocent between two women. Here is how a young woman puts it:

> Of course it's easier to have sexual relations with the same sex. Nobody even gets suspicious. They don't even get there in their imagination. I know it's hard to believe but you can't even imagine doing one percent of the things that you do with a woman with a man. When I was with her, we tried everything. Once we both got an orgasm in the subway car. Isn't that cool?

We showered together. We shared a bed. We had sex at my mother's place. We kissed in a park. Can I do these things with my boyfriend? Never! It's because no one can even imagine that we are in a relationship. They are busy with heterosexuality.

Findings of another study by this author (Arjmand and Ziari 2016) suggest that same-sex tendencies among young women is increasing, which is partially ascribed to socio-cultural constrains. When six young women were asked to compare their dating experience with a man in a gender-mixed park against same-sex dating in a women-only park, they all found dating the opposite sex to be "full of stress and apprehension."

You don't know what will happen in next five minutes. You could be arrested. A neighbor or a friend might see you and report you to the authorities or your family. And you have to go through the humiliation of a "virginity test." You are taking a big risk.

Iranian society is not yet there to imagine that those two young women holding hands are *lez* [a popular term for lesbians]. They understand it just as a sign of innocent affection rather than a relationship with sexual intentions.

This diverts and lifts the social pressure on same-sex dating and turns the parks into spaces re-appropriated for dating. To prove her point, a young woman tells a story of a covert lesbian (married under social and normative pressure) who was offered a lift by her husband to join her lesbian partner. On the surface, it looked like two course-mates were meeting to do their assignments and projects together, but in fact they were on a date. The same-sex relationship could continue without waking any sensitivity from families. The same woman compares it with the problems of meeting a partner from the opposite sex and concludes, "when I was dating a man, my family was not happy. They created all kinds of hassles to prevent me from dating him. But no one is even suspicious when I am dating a woman." She continues,

last week I had a quarrel with my mom and we ended up not talking to each other for days. My mom called my partner to come and talk to me and convince me to behave. For her we are good friends and she doesn't sense any danger in that. She is only concerned when a man appears in my life.

It seems that the segregation policy, despite its limiting effects in many aspects, provides space and opportunities for many women in different ways. They re-appropriate the space, which is created to maintain the Islamic rulings, to achieve the completely opposite effect and usage than originally intended. In the words of a 27-year-old woman,

Women-only parks are not the only places used for intimate same-sex relations, female dormitories are probably even better. Many start experiencing their first relationships from there. Relationships grow since authorities provide you with the space in which relationships are allowed and legitimate. Girls even shower together – maybe not necessarily with sexual intentions, though – but nobody ever labels them anything. I have always asked myself how one can interpret all this.

Hence, all rules and regulations are directed toward limiting the traditional heterosexual relationships and this in turn has resulted in a creation of a safe space, protected and enhanced by the policymakers and authorities, to be seized for same-sex sexual relationships in a safe environment without the harassments (verbal, sexual and otherwise) of men and of the police.

Families are among the groups of users who visit (gender-mixed) parks for recreational purposes most often. In the absence of green open spaces in modern Iranian houses – which were traditionally assumed as a part of any Persian home – many families, especially those in southern Tehran (where women-only Pardis Banvan and mixed Bi'sat Park are located), prefer to spend their weekend or other leisure time in the parks. Extended families, relatives and neighbors gather in the park, having a picnic and enjoying the fresh air and the vicinity to nature.

Regardless of the type of users, using fitness equipment, jogging and doing exercises in a green open space are among the main reasons for using the parks and are mentioned repeatedly by many users in both Niavaran and Bi'sat Parks. Patterns of usage vary among different groups of users depending on age, sex, personal interests and the like. While male users from various age ranges are most likely doing exercises, walking, jogging and using fitness equipment, most females prefer walking at a slow pace along the pathways. Many females, especially younger ones, noted that despite their will to use the fitness equipment they do not feel comfortable doing so in gender-mixed parks. "Nothing in law or religion would ban us from doing that, but you feel the heavy gaze of some men on you. It is not accepted by tradition," says a young woman.

A public space is one that is accessible and inclusive to all users, in the most convenient way and free of charge. This quality makes urban parks one of the most popular public spaces for family gatherings and group visits. Unlike the Niavaran neighborhood, where the neighboring residents have other possibilities (such as private gardens or farms on the city outskirts or a place at some attractive resort), Bi'sat Park in southern Tehran with its lower socio-economic surrounding functions as a recreational hub, providing a range of possibilities for the general public without any entrance or service fees. For families living in the neighboring districts, with rather limited means and resources at their disposal, a picnic at the park is one of the most favorite affordable activities on a weekend.

Vitality

The vitality of a public urban space is identified as the most important single factor that distinguishes a successful urban space from one that fails. It refers to the number of people in a given space over and across different times of the day and night. It also includes the uptake of amenities, number of occasional cultural events and celebrations, presence of an active street life in the vicinity of the space, and overall the extent to which a place feels alive and lively. As the definition maintains, the key to meeting the criteria of vitality within a public urban space is the presence of people, or as Montgomery (1998) puts it, "pedestrian flow" in and around it. Perceptibly, people tend to spend time in a place where they feel safe and comfortable, where they can find their favorite activities and where they enjoy being and spending time. Hence, higher population density and diversity in activities are regarded as prerequisites for the vitality of a public space.

The prime intention in building women-only parks – a public space devoted specifically to women – is to provide women with a space where they can perform and attend social activities without necessarily observing *hijab*. The objective has been to wake the interest of women to use such spaces and it seems to be a successful strategy for some women. To quote a young park visitor in Pardis Banvan:

> The most interesting thing for me, which made me eager to use the park more often, is biking. There are a few other parks in Tehran that women can bike. But this park is different. Here I don't have to cover my hair and put layers of cloths on. It's such an awesome feeling to ride the bike free without *hijab*.

Providing the possibility for unique experiences is likely to invite more people to the space and in turn may contribute to greater urban vitality. The uniqueness of the biking experience steers the preference of this woman to choose Pardis Banvan over all other options. The very unique experience associated with open-air biking is likely to bring her back to the park over and over again.

Vitality of a public space can also be studied in terms of social interaction among the users within it. Vital places are often used by a wide array of people (mixed users); this involves different groups using the same parts of a place; and different parts of a space provide venues for uses with different interests and preferences. In general, parks are types of public spaces that encourage people to meet, gather and socialize with each other. "It's not a place for feeling lonely. Look at all women around you here. They are laughing, chatting and eating. That is what makes the place vibrant and exciting," says a young woman in Pardis Banvan. One may argue that a coherent and supportive physical and functional structure provides more possibilities for social interaction in the form of various activities (including social, cultural, athletic, etc.) in a public space. Given those criteria, compared to the smaller women-only parks in Isfahan and Rasht, Pardis Banvan and Bihisht Madaran in Tehran have favorable conditions. The enormous size of the parks, larger spaces equipped with a wide range of amenities, activities and services that provide women with a series of recreational, educational and sport alternatives in Tehran, is in sharp contrast to women-only parks in Isfahan and Rasht with either modest or no such possibilities. During the observation sessions in Bihisht Madaran and Pardis Banvan a number of activities were carried out in different parts of the parks and women freely took part in them rather frequently. Variations in activities and the physical qualities of the space invite users to better experience the environment. A picnic with friends on the lawns, doing exercise in groups, participating in public gatherings and attending classes are among examples of such social interactions which take place regularly at those parks. Absence of such social interactions and activities in Isfahan and Rasht, however, could partially be attributed to the lack of amenities and services. Pardis Park in Isfahan is seldom used to its full capacity. Safety concerns mean women tend to use the park in small groups, rather than individually. Weak accessibility along with the lack of diverse activities resulted in the under-usage of Pardis Park. The small size of the park and its limited utilizable and furnished spaces also contributes to a limited interaction, which in turn affects the vitality of the place. Despite the good accessibility and excellent location, due to the lack of facilities and absence of activities in Park Banvan in Rasht, the park is rarely used by women and seems empty round-the-clock. "We come here with some friends, sitting, chatting, and can't do much more. I've never noticed any events or activities at this park. It's very small and nothing is in here to invite people," says a young woman visiting the park. The park only offers some sitting places in a green environment which composes the physical structure of the park and it is far from an ideal public place for social interaction and activities.

The long-term vitality of a public place can only be enhanced through a mix of uses and high pedestrian flow, both in terms of density and diversity. In cases where long-term vitality is out of grasp, planning occasional events and activities can help achieve short-term or periodical vitality. Temporary exhibitions, markets (seasonal and occasional), cultural and art festivals, religious celebrations and the like are instances of public events used successfully in some of the parks and could be emulated in others.

It is widely argued (Carmona 2003; Gehl 1989; Jacobs 1969; Montgomery 1998) that the co-occurrence of people in a space and time may provide opportunities for contacts and interaction (social and otherwise). The design of the park as a public space, hence, can create, inhibit or increase opportunities and enhance such encounters. Nevertheless, the space, the engagement, integrity and commitment of the people both in personal and collective levels – the extent they intend to be engaged in public life of the community – are decisive factors in creating such possibilities. Gehl (1996: 80–89) classifies various forms of contact from low intensity (passive contact) to enable one to be "among, to see, and to hear others," to high intensity (close friendship) which involves "emotional connections." He argues that what attracts people is other people:

> as opposed to being a passive observer of other people's experiences on television or video or film, in public spaces the individual himself is present, participating in a modest way, but most definitely participating. … Compared with experiencing buildings and other inanimate objects, experiencing people, who speak and move about, offers a wealth of sensual variation. No moment is like the previous or the following when people circulate among people.

(Gehl 1996: 83)

Carmona (2003) suggests the terms "active" and "passive engagement" to describe the extent of one's involvement in social life. While passive social engagement – in the form of watching people – may lead to a sense of comfort, active engagement requires a more direct experience with a place and the people within it. Some may find satisfaction in sitting close to the pedestrian flow or an ongoing activity, watching people while avoiding contact; some others may like more direct contact with friends, family or even strangers. Hence, "space is the system of relations" (Bourdieu 1989: 16) of various kinds and natures. Observations of different women-only parks suggest such different patterns of social engagements. In Bihisht Madaran and Pardis Banvan in Tehran one gets the impression that despite the parks being spaces conditioned by normative values, within the ramification of those values, they are spaces for active social engagements. The abundance of activities in those parks may provide an explanation for the tendency of women to make active social engagements in the parks. In the other women-only parks (in Isfahan and Rasht), however, a lack or scarcity of such activities has resulted in passive engagement. Various public activities in the form of permanent, temporary and occasional events will invite and attract more visitors to the park and provide linkage between people while it prompts interactions between people.

The process of social engagement is strongly influenced at various levels by the physical structure of a place and its design principles. A multi-purpose auditorium, classrooms, an open stage, and sitting places of various sizes, forms and shapes in Pardis Banvan are instances of a built environment which enhance the possibilities of public gatherings, social events and ultimately an active (and also passive) engagement of users.

Restrictive policies planned and implemented by the local authorities in Pardis Park in Isfahan, on the other hand, have contributed significantly to the discouragement of and, hence, lack of (social) interaction among users of the park. Regulations demand that women in groups of larger than three must register in advance to be able to utilize the park, a requirement that stands clearly against the principles relating to public places. Many women expressed their concerns about such regulations and argued that this (and similar policies) has further complicated the already problematic access and use of the park. The regulation requires planning in advance and a rather exhaustive process of application, registration and payment. The problem has not been left unnoticed by users, as one woman says, "the exhaustive process of deciding a time in advance, booking it, registering the day before and paying the entrance fee makes this park unattractive."

Interaction also appears in the form of networking, both "formal" and "informal." While formal networking takes place in the form of classes and courses of various types, informal networking is carried out among different groups of park users with similar interests. Urban studies scholars are attracted to the notion of informal network building in public spaces. These parks, in the absence of such possibilities through the institutions which make networking possible, are turned into a meeting place of various groups and create networks for different purposes, which often extend outside the parks. It indeed reflects the premise that people who are close together in social space tend to find themselves, by choice or by necessity, close to one another in geographical space; nevertheless, people who are very distant from each other in social space can encounter one another and interact, if only briefly and intermittently, in physical space

(Bourdieu 1989: 16)

In an informal public setting, any given person is in the position of both a decision-maker and a user, and interaction takes place simultaneously as one uses the space. The use of the same space and common interests facilitate the interaction, which is informal and spontaneous. Women who are interested in making connections, creating or expanding their networks with other women and trying to get to know others are familiar in women-only parks. The parks not only function as a much-needed place to provide a space for women to gather and socialize, but also triggers informal group building and networking. In an absence of social institutions and NGOs – which are restricted by the government – to promote social networks, women of similar interests such as young mothers with their babies, grandmothers with their grandchildren, and elderly women create their own networks and run their activities together. Informal networking is more common among the users of two women-only parks studied in Tehran compared to the other two in Isfahan and Rasht. Partial explanations could be the more vital social milieu and the presence of amenities that would attract users with similar interests. However, one should not underestimate the impact of two different lifestyles in these two contexts. While Tehrani park users are usually prone to a Western lifestyle with limited space at home and a nuclear family, the users from Isfahan and Rasht are more likely living a traditional life with extended family and already established networks. Most Tehranis, especially those in the southern part of the city, are migrants from other parts of the country and lack a family network in Tehran.

When visiting Niavaran Park, one comes to the realization that, in designing the park, due attention was paid to the vitality of the place as one of the main components of a successful urban space. The fact that the physical and functional structure of a space has a direct impact on its vitality was carefully considered in designing the park. A variety of usage and services at Niavaran Park along with its picturesque and pleasant environment has turned it into one of the most popular public spaces in the district. This has also fashioned a sense of belonging among the neighbors and users of the park and has made it a significant place for locals, as part of their individual and collective memories and an integrated and memorable part of everyday life in the Niavaran neighborhood.[3] Also, users from neighboring districts frequently visit the park and create their own social networks or join a group in one of the many gathering spots or *patuq*[4] at the park. *Patuq* could be interpreted as a form of Bourdieuian *habitus*, where "groups peruse strategies to produce and reproduce the conditions of their collective existence" (Hillier and Rooksby 2002: 398). *Patuq* fulfills an intriguing social function especially since gatherings of any kind, other than religious and normative-laden, meet resentment of the authorities and are restricted or banned.

Various functions attract a wide range of users to Niavaran Park at various times of the day and for different purposes. A well-known language education center (*Kanun-i Zaban-i Iran*), different playgrounds, sport center, café and out-door sport facilities (especially a popular roller rink) are examples of such spaces and functions that invite many users, especially teenagers and young adults, to the park. Many others, particularly elderly users, visit the park on a regular basis. For this group of users the park is a place where they walk, exercise and meet their friends and neighbors. They usually belong to various networks at the park. Many know each other and take a moment to greet each other while walking or strolling at the park. Such interactions seem to be a result of and enhanced by the continu-ous and regular presence and social networking at the park.

People enjoy their favorite activities in a place where they feel safe and com-fortable. A responsive physical structure will encourage more people to use the space and, hence, improve its vitality not least through increasing the population density, flow and public presence at the park. A good example is the plane area, surrounding the main water basin, where the physical characteristics of the space have turned it into one of the most popular and vital spots at Niavaran Park. The area is well furnished, properly equipped with various training facilities, has a nice view over the surrounding greenery and beyond, lacks slope and enjoys a good view over the main water basin. All these characteristics have made this part of the park a spot preferred by many users.

Pedestrian flow not only promotes the vitality of a space but also improves the sense of safety and invites more users into the park. A wide range of users – both women and men of various ages, except teenagers and young adults – prefer using the crowded part of the park. The presence of women, exercising or doing differ-ent activities in various parts of the park is what makes Niavaran Park different from all others. Women in brightly colored sport suits, mostly pretentiously dem-onstrating exclusive Western brands[5] (despite all restrictions), are among those who walk and exercise, alone or in small groups, at Niavaran Park. The presence of women using the space in such a free manner contributes to the vitality of the space and is an indication of safety and security in the park. When asked about "feeling safe" a young woman notes: "Sad but true. We feel safer when the police are not around. We are more harassed by the police for the way we look and dress than by those whom the police are trying to protect us from."

The southwest corner of the park, equipped with new fitness equipment, is another popular spot in Niavaran Park which attracts many users. A healthy life-style through training the body seems to be a serious concern of the local residents in this part of Tehran and the park provides them a venue for those activities. With their picturesque and pleasant green spaces in the midst of an otherwise dense urban fabric, the parks are among the public spaces that encourage and attract people to meet, gather and socialize with others. Niavaran Park is a prime instance of such a popular urban park in Tehran, attracting a wide range of users to ben-efit from the park for a variety of purposes. This multi-facet and multi-purpose function has greatly contributed to the improvement of the quality of public life within Niavaran and the neighboring districts. The park is used as *patuq* by many different groups of people for a variety of purposes.

An urban space is successful in its function when it promotes social interaction and facilitates communication among people. The design of the physical space affects and improves social interaction among people by providing subspaces where people – of various parts of the urban fabric and diverse backgrounds – get together naturally and interact without any obstruction. The arrangement of furniture in a certain manner can also enhance and promote social interactions, remove barriers among people, form a friendly environment and set the stage for interaction.

In Niavaran Park, it is not solely the design and arrangement of the furniture but also subspaces such as the stairways and cozy corners that provide sitting and gathering spots for the users. Taking the climatic conditions into consideration in designing the park, such subspaces resulted in the creation of comfortable venues for various purposes that could be used at different times of the year. Taking a moment under the shady trees and feeling the cool breeze that passes through on Tehran's hot summer days is an example of how such spots encourage people to enjoy the park. It is not unusual to see a group of young (or older) women, or a group of younger women and men together in one of the many cozy corners of the park. It is a more accepted practice for women in Niavaran to interact and be visible in public than in other parts of Tehran.

Unlike the physical space of Niavaran Park, which strongly supports social interaction and promotes friendly gatherings among users, the arrangement of the furniture at Bi'sat Park has a more distancing effect on visitors.[6]

Despite the fact that all benches are located under the trees and enjoy the shady spaces in summer, they are situated sporadically, separately and scattered. The distance between benches and their arrangement – in a row and not in facing each other – provides little or no possibility for people to sit in a group or form a friendly gathering. The lack of appropriate arrangement of furniture made users at Bi'sat Park adopt various strategies to create a better space for interaction. A group of elderly men who visit the park every morning abandoned the benches to sit on the two parallel short walls located in a corner of the park. Facing each other, the walls provide a good possibility for gathering and interaction. However, such a possibility is much more limited for women.

While most users of Niavaran Park are locals who live in the neighboring districts, Bi'sat Park receives a wider range of people who use the park for various purposes. The diversity among the users of Bi'sat Park is strongly affected by the neighboring functions. In addition to the surrounding residential areas from which the main groups of local users visit the park; Payam Noor University campus (a partially virtual university with campuses around the country) at the northwest corner of the park, two schools in the south and a cross-country bus terminal in the west are other functions which introduce new groups of users to the park. Contrary to the users of Niavaran Park, however, users of Bi'sat Park seldom develop any sense of belonging to the park.

Despite the fact that the Bi'sat Park is very large and provides many facilities and services for its users, the local people do not demonstrate enthusiasm to use them on a regular basis. Using a park for sports and exercise – one of the common activities at Niavaran Park – is not equally common in Bi'sat Park. There are, however, certain groups of people, such as the elderly, who visit the park frequently for social networking. Local residents mainly use the park for extended family gatherings and weekend picnics.

The South Cross-Country Bus Terminal – one of the main transportation hubs of the Tehran metropolis – introduces new users to the park on a daily basis. Many passengers entering the terminal from different parts of the country or who are leaving Tehran use the park as the waiting room: a place to rest, to meet a friend or to wait for their bus to arrive. For this group the park is not solely a place to enjoy the green environment or activities within it, but is the only available alternative in the neighborhood to spend some time while waiting. Many people traveling from smaller cities find refuge in the green milieu of the park in the midst of a city of heat, pollution and noise.

To see passengers with their suitcases and bags taking a nap in the shade or on the lawn is a familiar scene at the Bi'sat Park. A young woman at the park says,

> I'm coming from Burujird. I arrived at half past five in the morning (it is 09.45 now). I went to the praying room at the terminal to sleep for a few hours, but they closed it after the morning prayer. So I came here. I'm waiting for one of my relatives to come and pick me up. I don't know Tehran well. I have to wait for him to come.

A female student (with a young man introduced as her boyfriend escorting her) is returning back to her hometown for a holiday:

> We arrived at the terminal too early and decided to spend our waiting time here in the park. It is much more pleasant here than in the terminal waiting hall.

The vitality of a public space is strongly affected by social interaction among its users. Urban parks turn into vital places if they enjoy a combination of various users. The start of the academic year in autumn introduces a new life to Bi'sat Park by increasing the number of students using the park exponentially. Many of them use the park after or between their classes as a place for dating or meeting up with their friends. Vicinity to the university has enhanced the diversity of young groups of users at the park. An overrepresentation of students who are not necessarily living in southern Tehran has resulted in a greater mix among the park users. Their appearance – i.e. the way they dress – and patterns of usage are different from other groups using the park and their presence lends a sense of vitality to the park. Lack of appropriate space for socialization on the campus, constant presence of security guards and moral police, strict mechanisms of control and surveillance in the university are among the reasons that encourage students to spend their free time at the park. Although the park is also under the surveillance by the police and security guards, they are not as strict and restrictive as those at the university. Students, hence, can interact and socialize more freely with their peers and spend time in a more pleasant environment. This is seen equally among both female and male students, although the pattern of usage is different among female students compared to males (discussed more thoroughly under the headings, "a place for all" and "safety and security").

Despite the fact that the presence of a wide range of users is more visible at Bi'sat Park, Niavaran Park does enjoy greater vitality. The total area of Bi'sat Park is some nine times larger than Niavaran Park. The size of the park and hence the lesser usage of the space are among the factors contributing to a lesser or lack of vitality. One may also observe other reasons for it:

- The park stands alone in the midst of its surrounding environment and lacks linkage to the nearby urban fabric.
- The park lacks compatibility with the neighboring functions.
- Despite the size, the park lacks enough supportive and joyful activities.
- There are other parks available in the same part of the city.
- It is perceived that there is a lack of security in the park.

The third reason is a fact noted by many users and pointed out repeatedly. However, this was when the amusement complex inside Bi'sat Park was closed. Many mentioned that as a reason for the unattractiveness of the park. Some others referred to privatization of the amusement park and argued that it had not resulted in the effects and efficiency expected. Using the amusement park became expensive and not everybody in that part of the city could afford it, and the equipment was not maintained properly. Using worn-out equipment, which does not meet safety standards, is a concern raised by many users. One woman from the neighborhood thinks that lack of safety is one of the main reasons for people losing interest in visiting the park:

> I remember the time when the amusement park was open and things were in good shape. Many families used to come to the park at weekends and the park was always crowded. Despite the huge parking lot, people could hardly find room to park their cars to use the park.

A woman points out the lack of attractive and joyful activities at the park after closure of the amusement park:

> This park is within a stone's throw of my house and we used to come here quite often. Since the equipment broke and the amusement park closed, we've seldom been here at the park. I remember the time when my husband and I used to come to the park at least 3 or 4 nights a week.

Bi'sat Park, as the only public park in southern Tehran for many years, has served as one of the most popular public spaces for inhabitants from the southern part of the metropolis. Recently, however, the municipality has launched new neighborhood public parks in this part of the city. Diversity in themes, variations in size and location and an abundance of facilities and services in those parks provide more alternatives for people to choose their favorite public spaces, and this affects the frequency and pattern of their usage of Bi'sat Park.

Lack of a sense of security inside Bi'sat Park is also mentioned as a reason that the park is less inviting for people and, hence, less vital. The park carries a reputation for being a place for the homeless, drug addicts and dealers. Despite the heavy presence of police and lesser visibility of such groups at the park, the image of the park as an insecure and unpleasant place, is still alive in people's mind. The presence of police and security guards, which started in 2010, was part of the effort to provide a safer environment. This, along with the surveillance systems has been part of the endeavor to increase the sense of safety and security inside the park and invite more users to it. Although those measures have been successful to a certain extent in providing a sense of security and safety, they have also affected the presence of younger users who perceive that such measures limit their freedom in the park. A heavy presence of police and security guards, argued to safeguard the park for users, has had a significant negative effect on the vitality of the park. The normative-laden regulations, prompted and reminded constantly by the strolling police, have limited the sense of freedom and affected the presence of users and ultimately the vitality of the park.

Safety and security

A successful urban space is expected to be associated with a sense of security and safety,[7] as an essential constituent of such spaces. Carmona (2003: 119–20) argues that security relates to the "protection" of oneself, one's family and friends and properties (both individual and communal). Lack of security, feeling in danger and a fear of being victimized threaten both the use of the public spaces and the creation of successful urban spaces. Also, as Lawson (2001) argues, a sense of security is a profound and fundamental requirement for creating a sense of stability, sustainability, continuity and predictability in one's social life.

While the notion of safety and security is undoubtedly a pivotal factor in creating a successful space, it is important to make a distinction between "fear" and "risk"; this is referred to as the difference between "feeling safe," for instance due to the reputation of a place, and "actually being safe," i.e. being safe despite a reputation which may indicate the contrary. It is crucial to take this into consideration, since women are more vulnerable and a more at-risk group in society compared to men, and hence prone to feel more in danger and be subject to victimization. Feeling in constant fear is a result of a "burden of female body" (Brownmiller 1975: 15–17) and is an experience shared by many women across various cultures, rooted in the dominance of the patriarchal structure in society. To avoid unnecessary risks, women often take cautious measures and may adopt a conservative social life to minimize the risk. This would include avoiding given public places and activities which may put them in danger. While observing certain design principles can facilitate building a safer space (physically and otherwise), those principles are, however, compromised in designing women-only parks, where seclusion and enclosure are of prime importance. Those exclusive features in the design of the women-only parks contribute to the reproduction and maintenance of the symbolic violence where the segregated space "imposes meanings as legitimate by concealing the power relations which are the basis of its force, adds its own specifically symbolic force to those power relations" (Bourdieu and Passeron 1990: 4). Creating a gated space to exclude males, adding extra physical protection and improving the surveillance systems inside the park are among the endeavors that try to lend a sense of safety to park users and hence legitimize such spaces as a solution to safety issues across society at large. Despite the mixed reflections on the issue, many women seem to be satisfied with a segregated space as a solution. In the words of a young woman in Pardis Park,

> As a woman I feel safe here in the park and no one bothers me. The women-only homogenous environment gives a feeling of safety and comfort. No one is gazing at me here, which makes me feel uncomfortable in regular parks.

Another woman in Pardis Banvan shows her satisfaction of the safe environment inside the park in the following way:

> I bring my daughters here to skate. Here is the only place they can skate without any need to observe *hijab*. It's really safe in here and I'm not worried about them. In the ordinary parks out there I never leave them out of my sight. Here I can enjoy a moment for myself too.

Gated space along with additional protective means and strategies has increased the sense of security inside the parks. However, the parks are situated in a larger urban context which makes it impossible to apply the same means and strategies outside or around them. It also creates a paradoxical situation. The primacy of privacy and restrictive normative on visibility has forced the authorities to build these parks in marginal areas or on the outskirts of cities, an unsafe place by default compared to the inner parts of the cities.

The problem is more visible and persistent in Pardis Banvan in Tehran and Pardis Park in Isfahan. The demographics of these two parks have created a number of issues for the users, which are noted repeatedly. Complaining about the safety, a young woman in Pardis Banvan notes,

> I don't feel safe coming to the park alone. I should always ask my husband to give me a lift. The park closes too early in the evening and I have to wait for my husband to return from work to drive me home. Not practical at all. With this, the only choice is for me to come to the park on Fridays, which is also the only day that my husband and the rest of the family are home together and I prefer to stay home. So, even if I like coming, practically I can't.

Feeling unsafe, mainly due to the location of the park forces her to be dependent on her husband, against the fundamental notion for the creation of those parks. Another young park user referring to her stressful experience of coming to the park for the first time says,

> Not only is the park located too far from where I live, but it is in an unsafe area too. I was nervous coming here. I came with my friend and none of us knew this area well. We had always heard that southern Tehran is different from other parts of the city. We were told that it is not safe.

A glance over the aerial photo of Pardis Banvan and its surroundings confirms the concerns of the park users and their feeling of being unsafe. The park is clearly isolated from the urban fabric by surrounding highways and neighboring functions. Lack of a functioning public transportation network, small-scale businesses, residential blocks and any other such amenities and activities has intensified the problem and contributed to the safety and security issues. People tend to feel safer where the social life takes place and where a place looks socially and functionally vital.

Pardis Park in Isfahan, located at the western outskirts of the city and surrounded by a highway in north and farmlands in the south, suffers from the same problem. As a result of cumbersome rules and regulations on women-only parks not being visually accessible from buildings nearby, the park has remained completely isolated.

Observations from the Bihisht Madaran Park in Tehran and Park Banvan in Rasht supported by users' responses demonstrate how choosing an appropriate location and connectivity to other parts of the city can increase the number of visitors and ultimately improve the safety of a public place. Interviews with park users reveal that the notion of safety is so important to the users that women from different parts of the city undertake the hassle of traveling longer to use a park in safer surroundings. Visitors from southern Tehran, despite being closer to other local women-only parks, prefer to use Bihisht Madaran for safety and security reasons. Time and again, women expressed that Bihisht Madaran is their first choice as it is located in a busy district of the city with good connectivity and access to the public transportation network, and because it is situated in a safer district compared to all other women-only parks in Tehran.

Feeling unsafe and in danger and fear of victimization affects the presence of women in a public space. Feeling unsafe in a place could be partly an outcome of an unresponsive and inappropriate physical and functional structure of a space. Principles of space planning and design can strongly contribute to the sense of security and safety among users. For instance, creating *indefensible spaces* – spaces in which a person perceives as not capable of being defended – in urban parks and surrounding areas can make people feel unsafe and in fear. Urban design principals could be utilized to promote the defensibility of a space through natural, formal and mechanical surveillance measures. It is a commonly agreed account that, when offenders get the sense that they could be seen by others (even if they are not) they are less likely to commit an offense in such space.

Hence, compared to the users of women-only parks the users of Niavaran and Bi'sat Parks experience a different sense of security and safety inside the parks. In general, users of Niavaran Park, despite a lack of hard surveillance systems and presence of police, perceive the park as a safe place. The constant presence of people in the park, reinforced by the variety of activities within the space, provides natural surveillance possibilities and enhances the sense of security and safety among users. Adapting appropriate design strategies such as improving the visibility of places and providing additional lighting have strengthened the natural surveillance, reduced the sense of fear and ultimately improved the feeling of security inside the park.

Unlike Niavaran Park, however, users of Bi'sat Park perceive it as a public space with a lower security and safety. The park has earned a reputation as a place occupied by drug addicts, drug dealers and the homeless who may harass or offend others. Families living in the neighborhood show concern about letting their children and young adults spend time in the park.

Regular surveillance by police patrols has enhanced the security of Bi'sat Park in recent years. This, to some extent, has improved the users' image and perception of the park. A police station inside the park and the constant presence of security guards round-the-clock is one of the main differences between Bi'sat and Niavaran Parks. The security guards identify and register all criminal activities and suspicious behavior to prevent offenses and promote the image of the park in the neighborhood.

Many users have stories about sale of alcohol and drugs in specific spots at the Bi'sat Park before the presence of police and guards. A young man, pointing at the old airplane in the park says,

> You see the plane over there. On some occasions, you could extract about a full container of drugs from underneath the plane. I've seen it with my own eyes. My friends and I knew that if we needed any booze for parties, we would find some there.

In addition to the physical structure and social context of the park, the functional character of the spaces, both in micro (inside the park) and meso scales (in the surrounding fabric), is a major factor in promoting the sense of security among users. In Bi'sat Park, it seems like an ignored fact that the sense of security invested by the police patrols could be promoted by a series of supportive activities to invite more users into the park. Supportive activities here refer to design features of the space to provide intended activities within the space as well as providing the possibility of a mixed-use of the park as a public urban space.

Bi'sat Park also suffers from a lack of supportive activities: the amusement park has been closed for a number of years, the food corner is only working a few hours in the afternoon, and the open stage and amphitheater is rarely used due to the lack of public events scheduled in the park. Creating a mix of activities would increase the number of users in the park and provide presence of the public in the space, which in turn would result in denying potential opportunities for crime and offensive behavior.

Sense of security and safety varies among different users and differs from one group to another. Sense of security is one of the areas in which a significant difference between female and male users is observed. The differences include the type of subspaces used for given activities and the time of the day they prefer to use the park. Vicinity to the university to a certain extent has compensated the lack of users at the Bi'sat Park and has created the opportunity for students to enjoy the park in the absence of a proper educational space on campus. Female students tend to adopt certain strategies to avoid potential risks at the park. Many female students prefer to spend time at the park after or in between their classes, but often leave the park before dark. They seem to tacitly acknowledge the dominance of "orthodox male sexuality" which affects women through a "system of male intimidation that keeps women in fear … [which] provokes the need to be much more cautious in everyday aspects of life than men have to be" (Brownmiller 1975: 15). All women interviewed at Bi'sat Park noted that they do not feel safe being at the park after dark and they all are in constant fear of harassment by men, mostly drug addicts or homeless people. While verbal and sometimes physical harassment is a rather common problem in Iranian society in general, when they were asked if they had ever been exposed to harassment by drug dealers or criminals at the park, it seemed that they had always heard "someone" has been harassed. It seems to be more of a feeling, not substantiated by any personal experience or fact. It is more of a reputation than actual fact. While showing concerns about offensive acts, students also found the park a place where they can interact more freely with their peers, away from the watching eyes of the security apparatus at the university. They mostly prefer visible and crowded parts of the park as their *patuqs* (gathering spots). Unlike the male users of Bi'sat Park, women select *patuqs* close to main paths of the park, spots not so private and hence unpleasant. This is a precautious measure to avoid any undesirable encounters with men and stems from feeling unsafe and a fear of offenses of a sexual nature. The sense of fear and feeling there is a lack of safety relates to what Giddens (2009: 965) explains as "ties between sexuality and [masculine] power and superiority" in society at large. "There is a sense in which all women are victims of rape. Women who have never been raped often experience anxieties similar to those who have" (*ibid.*).

Moreover, in addition to the park itself, the character of the outer space and the surroundings lends itself to a sense of safety and security among users. The question of safety and security, however, is tangled with the notion of surveillance.

[I]ts predominant strategic function is the creation of a space for a "subject people" through the production of knowledges in terms of which surveillance is exercised and a complex form of pleasure/unpleasure is incited. It seeks authorization for its strategies.

(Bhabha 1994: 70)

On the other hand, while historical and cultural functions along with the higher socio-economic community have promoted a sense of security in the Bihisht Madaran and Niavaran neighborhoods, variations of the functions both in terms of type and scale, have provided a mix of incompatible activities in the area where the Pardis Banvan, Pardis and Bi'sat Parks are located. While Bihisht Madaran and Niavaran Parks are visibly dominated by users of the neighboring residential areas and visitors from different parts of the city arguably seeking a safer space, Bi'sat Park hosts a wide range of temporary users; very few from the neighborhood. The lack of a sense of belonging among such users reduces their responsibility and concern about the park as a local public space of one's own community. In addition to the diversity of users, the large size and the organic spatial design have created a number of indefensible spaces which reduce the sense of security and safety inside the park.

Sense of place

Sense of place is a key concept in urban design and suggests that places are not solely physical structures but entities that encompass meanings which people associate with and assign to them. The term "meaning" here refers to and includes the perception of people from a place, their experiences, feelings and memories. The way people recognize a place is not just by its physical appearance but through its functions, human interactions and the like. Hence, for the same location, the sense of place will vary from one person to another and for the same person over time. Improving the sense of place has always been one of the main goals of creating a successful public space and as a multidimensional construct has been defined in a variety of ways. According to Eisenhauer *et al.* (2000: 422) sense of place refers to "the connections people have with the land, their perceptions of the relationships between themselves and a place, and is a concept that encompasses symbolic and emotional aspects." Billig (2005: 118) describes sense of place as "the atmosphere to a place, the quality of its environment and possibly its attraction by causing a certain indefinable sense of well-being that makes people [want] to return to that place." Hence, sense of place is a multidimensional and complex construct which characterizes the relationship between people and place and is rooted in behavioral commitments; it determines how people perceive the environment and emotionally respond to it. Sense of place is perceived through one's habitus, "the way of knowing the world, a set of divisions of space and time, of people and things which structure social practice" (Bourdieu 1990: 210). Bourdieu utilizes "structuring structure" toward the ways in which the habitus shapes social practice and sense of place provides the grounds for reproduction of social practices conditioned by the previous experience of the space. Sense of place is also affected by the surrounding social realities. In such context as Iran where the female body in public is restricted by law and women are subject to what Foucault calls "institutionalized bodily discipline," the sense of place in a space occupied entirely by women is likely to differ to that in gender-mixed spaces. As Jorgensen and Stedman (2006: 316) argue, sense of place "represent[s] beliefs, emotions, and behavioral commitments concerning a particular geographic setting." Such construct, however, is shaped in part by the power dynamics of society. In the process of constructing a sense of place in everyday practices, an individual selects a series of creative methods or "tactics" in the battlefield of everyday life. By their very nature tactics are defensive and opportunistic, used in more limited ways and seized momentarily within spaces, both physical and psychological, and produced and governed by more powerful strategic relations (de Certeau 1984).

Despite the differences in scope and disciplinary variations, what all those definitions agree upon is the significance of the sense of place as a key constituent in creating a successful urban space. To reach a sense of place three essential components are combined and realized: a physical space, an activity(ies) and the sensory experience(s). As noted through the analyses of data (interviews and observations), there are two distinct trends in women's sense of place and their perception of the women-only parks (discussed in detail in Chapter 5).

The first trend involves a group of women who regularly use the parks and perceive them as a gesture of respect to women in a male-dominated society. When they were asked about the notion of gender segregation in society, however, the group split into two subgroups. The first subgroup consists of (mostly practicing) Muslim women with strong religious ties who accept the normative religious-laden values of the Iranian society and consider the segregation of sexes in public legitimate and an inseparable component of Muslim society. They are not necessarily supporters of the Iranian Islamic government *per se*; however, many of them do not see any problem with issues such as compulsory veiling and segregated spaces in such an Islamic society as Iran. The second subgroup consists of women who are not necessarily religious and who demonstrate resentment against the segregation of sexes in society. However, they reckon these parks as an open space for women-only activities in an otherwise male-dominated society. For them, these parks provide the possibility of activities that are not available in the ordinary parks. A frequent user of Bihisht Madaran, who is in her thirties, elaborates on this:

> I like to work out and in other parks I feel that everybody is gazing at me. It makes me uncomfortable and I have to avoid those parts of the workout that may attract the attention of others. In these parks I don't need to cover my body and I can do any exercise without feeling uncomfortable.

This group, while using the parks (and sometimes more frequently than others), express strong opinions about the parks. Often they make critical comments on the structure, functions and facilities, and management, and sometimes provide ideas and solutions to improve the quality of the parks and their services.

The second trend involves a group of women who actively resent the idea of segregated parks for women and seldom (or never) use the parks and perceive them as yet another instrument of institutionalization of gender segregation in an already highly-segregated society. While the former group criticizes the physical and functional features of the parks, this group focuses on the socio-psychological aspects of the parks. For this group, society encourages the separation of sexes and institutionalizes it as a norm. Children at schools, young adults in their leisure activities, women and men in most levels of society are segregated and there is no chance to learn to interact and communicate with each other. Many in this group acknowledge that, in such a closed society as Iran's, these parks are windows of opportunity and possibility for women, but they also argue that the institutionalized separation will harm society in the long run. Tahmineh Milani (an architect by training), an acclaimed director and a public figure, puts it this way:

> this is not a solution, maybe just a temporary remedy ... at some point, women and men are expected to work together and if they have not been exposed to each other and have not learned how to interact, what would happen?
>
> (IRIB 2013)

This outlook is shared by many feminists and women's rights activists.

Although this dichotomy makes a clear distinction between the first group as religious-minded and the second group as non-religious, one should not essentialize religion as a decisive factor in shaping their views. The sense of place is most likely perceived and maintained by the physical structure, activities and personal experiences or backgrounds of the users. Many women vividly remembered their first visit to a women-only park as an exciting experience, "like nothing I had experienced before." The uniqueness of such an experience makes it memorable and interesting for the user. A young woman in Pardis Banvan expresses this in the following way:

> It's such a nice place and quite a new experience for me. I was born after the revolution. I've never seen women without *hijab* outside the family. Not even in my all-girls school. It's the first time I have ever seen women not wearing *hijab* in a public space ... Such a wonderful scene ...

Many women express feelings that indicate the uniqueness of the experience associated with women-only parks. They remember their first encounter with the parks as the thing that encouraged them to further visit the parks, as a sense of place which could not be replicated anywhere else in the country. In the words of young woman in Bihisht Madaran,

> The sense of freedom that women experience here is not comparable to anything else and can't be experienced anywhere else in Iran. I take it as a kind of respect that the municipality devotes such parks and their services and facilities only to women. ... When I came here for the first time, I found it very comfortable and friendly. It feels really good in here and it's safe to be here: sitting and chatting with friends, having lunch together, walking, etc. ... We laugh loud and nobody cares if she is heard.

A sense of place is gained when a user expresses her satisfaction of the place. Attractiveness, variations in physical design, functions and activities as well as access to different amenities positively affect the sense of place among users. Pardis Banvan and Bihisht Madaran (both in Tehran) are good instances of women-only parks with numerous possibilities. Variations in physical and functional possibilities provide the grounds for women to get involved in a variety of social and cultural events, educational programs, sport activities, informal gatherings, networking and the like. Higher satisfaction of the users of these parks is likely to be enhanced by these amenities, which in turn affects the users' sense of place. In contrast, Pardis Park in Isfahan and Park Banvan in Rasht with less or no such possibilities lack the uniqueness of the experience and result in a not-so-remarkable experience of a place. A weaker sense of place does not motivate the user to return and utilize or replicate the experience of the park again.

Sense of place is also achieved by the personalization of the place. Human beings are always in need of a sense of identity, of belonging to a specific group, community and territory. A place can provide a physical context for shared experiences and collective memories among people who are attached to it. "A sense of belonging may be achieved by physical separation or distinctiveness, and/or a sense of entering into a particular area" (Clifford and King 1993: 97). This physical separation – creating boundaries around the space and defining "insiders" and "outsiders" – refers to "territoriality." Thomas and Ahmad (2004: 12) highlight the bonds that unite people rather than the differences that separate them: "The factors that make communities cohesive are complex. ... Good design and place management can contribute to a more widespread sense of belonging and can foster good relations between and within communities."

Awareness and goal-centeredness in planning, design, performance and management of a public space can encourage successful integration and contribute to building cohesive communities. Cohesive communities are characterized by their willingness to participate in decision-making and implementation of whatever relates to the shared spaces. Such communities also affect the way people experience places as friendly and welcoming and foster a sense of belonging among their members.

Women-only parks in Iran can be considered an example of defining a territory which separates the female insiders from the world of the outsiders through physical boundaries. Women, despite their mixed social, cultural and economic backgrounds, are members of this territory. This sense of belonging, while positive on one hand, is also undesirable on the other. It helps to construct a male "other" and widens the gender gap further.

Sense of place is gained through particular experiences, feelings and memories of a given space. Various components of a space including physical structures and functional layout, along with the human dimension – users and their interaction with the space – shape the sense of a place among users. Hence, the sense of place is not solely about the space and its components, rather the personal traits of the users are the decisive factor. Despite the fact that the sense of place is subjective and varies from one person to next, there are similarities among different groups of users depending on their cultural affiliations, socio-economic background and interests. Moreover, the characteristics of the space and its functioning features can also contribute to create a similar image and perception among different users.

Urban parks are places to relax and refresh in the midst of the crowded, polluted and noisy urban fabric surrounding them. Many users at both Niavaran and Bi'sat Parks repeatedly mentioned the "trees," "greenery," "flowers," "colorful composition of natural elements," "fresh air," "water" and "sounds of birds and running water" as the main components to turn the urban parks into beautiful and attractive microclimates. The contrast of those parks with their surrounding urban areas is mentioned as the main factor in allowing people to relax their minds from the concerns of daily life, perusing joy and pleasure in a green environment. The beautiful natural environment of the parks results in mental balance and creates a positive mood among the users. A young women visiting Niavaran Park with her family notes,

> The park makes me feel happy. In the routine and robotic life of today, it gives a different experience. We go to work in the morning and come back in the evening, every single day. Coming to the park is a change in the routine. It uplifts my mood.

Families are among the main groups of users who visit the parks to enjoy the greenery and beauty of the environment. Living in such a large city as Tehran, families find the green environment of the parks a refuge in which to spend a pleasant and joyful time. "A picnic with family on a weekend," "a dinner in the evening in the open air," "a moment playing with children on the lawn" or "being part of a public event held in the vast green area" are among the main purposes that bring families into the parks. While women-only parks do not provide the possibility for the entire family to spend time together at the park, many users at Niavaran and Bi'sat mixed parks – from the neighboring districts, usually in walking distance and easy access of the parks – visit the parks with their families.

Groups of families, sitting on the lawns, chatting, playing and eating – especially on the weekends – are familiar scenes at parks. A women who is part of a large group of several extended families at Niavaran Park argues,

> I really feel good when I come to the park. Being with the family and spending time together is very joyful for me. Every time, when we are leaving we make a plan to meet up here again soon. This is only possible in mixed parks.

Residences of the small tenements seem to enjoy the parks as a playground for their children. They often mention the quality of the air and green environment of the park as well as the playgrounds that provide a good possibility for children to play and enjoy. Parents can do their own activities while spending time with their children. A young mother at Bi'sat Park says,

> My kids like the park a lot. They can play here freely and loud and we are not worried about the noise they are making. When we are at home, we usually

gaze at the TV without a word to each other, but here we talk to each other and play with the kids. It's quality time for the family.

The attractiveness of the green environment for users is improved by adding extra amenities, services and the possibility of further activities targeting different groups at the parks. In addition to the visual attractiveness of the parks, fitness equipment and playground facilities attract many users to the parks. A successful example is the chessboard table at Niavaran Park. Playing chess seems to be one of the most popular activities at the park since users, especially the elderly, are often busy playing. Despite the fact that people can play at home or use fitness equipment in gyms or sport centers they consider the openness of the space as an uncompromising privilege. A woman at Niavaran Park says,

> Nothing is better than doing exercise in a place surrounded by the natural beauty of the park. I have some of this equipment at home, though. But I prefer to use them here. Besides, I've heard that it's better to do such exercises in the open air where you can inhale the fresh air. That's why I come to the park. Even though I suspect some equipment – especially the older equipment – doesn't meet the standard quality.

In addition to the green and natural character of the parks, the publicness of such spaces turns them into one of the most popular urban spaces in Tehran. Having a wide range of users, urban parks promote social interaction among users and enhance the public life of the community. Various groups of users use the parks as gathering spots regularly. At both Niavaran and Bi'sat Parks, the elderly are among the frequent users visiting and using the parks for various purposes. Living in such a metropolis, many older people are lonely and find themselves isolated in their – usually small – tenements in southern Tehran. For such users parks are safe and beautiful green spaces, which provide opportunities for various activities. By spending a few hours a day at the park, they meet new people, create and maintain their social networks and keep their social engagement with society, while walking, enjoying the beauty of natural environment, doing exercise and attending social events to improve mental and physical health. Judging by the frequency of visits, Niavaran Park seems to be a successful example of such spaces. It has turned many locals in the neighborhood into frequent users, fond of the pleasant and lively environment of the park. They consider the park as one of the main public spaces in the area and a major public space to facilitate social life in the Niavaran neighborhood. Such frequent visitors – mostly elderly males – have their favorite spots and meet with a specific group of friends, which not only gives them the feeling of making one's own space (*patuq*) but also a sense of belonging to the space. Older women, however, do not visit the park as frequently as men do and rarely stick to the same spot. An older woman in Niavaran Park says,

I like this park. From a landscape architecture point of view, it's one of the most beautiful parks in Tehran. Despite the fact that it was designed and built many years ago, it still holds the very character and identity…

There are groups of elderly men who visit the park and meet their friends on regular basis (and we met them every time we visited the parks). Sitting and chatting with friends is an opportunity for them as retired people – with few or no possibilities – to be engaged in the social life of the city. A 78-year-old man sitting in a group of elderly men at Niavaran Park explains,

I'm so used to this place. If I stay home and don't come to the park, I would feel down for the entire day. I've got old friends here. Other parks might be more beautiful but my friends are not there. Every day when I come here, I meet my friends and talk to them and I feel at peace. That's what I like most about my friends and Niavaran Park.

There is, however, a notable difference between women and men in the pattern of usage of the parks. While group gatherings and social interactions seem to be very popular among many men, many women prefer to enjoy the quiet and peaceful corners of the park. Women usually come alone or in the company of a small group of two or three, try to avoid noisy spots and group activities, and almost never form a *patuq* similar to men's. Doing light exercises, walking, sitting on benches, reading books or newspapers and talking with a friend are among the favorite activities of this group of users. The female body in public space has become the subject of power discourse, hence, in de Certeau's (1984: 93) terms, it

follow[s] the thicks and thins of the urban text … the networks of these moving intersecting writings compose a manifold story that has neither author nor spectator, shaped out of fragments of trajectories and alterations of spaces in related to representations, it remains daily and indefinitely other.

Notes

1 The urban fortress model is often used to define those gated communities established mostly to control crime. They include enclosed neighborhoods that have controlled access through gates or booms across existing roads, and security villages and complexes, including lifestyle communities which provide their enclosed residents with a range of non-residential amenities such as schools, offices, shops and golf courses (Landman and Schönteich 2002: 71–83). Although it remains as one of the popular methods of crime prevention for those who can afford it, many scholars warn of the long-term impact of such isolation on society at large. Such arguments are also raised against women-only parks, which will be discussed in Chapter 5.
2 The term is used by the Iranian government and propagated systematically through state media to include Western cultural products, which are perceived to be leading an organized "soft war" against Islamic values. Using this argument, the government initiated a campaign against Western "cultural assault" and media imperialism, in what it perceived as the West's efforts to counter and dismantle the regime by "soft power."

The proponents of the cultural assault thesis claimed that the West, particularly the United States, "having been disappointed in toppling the Islamic Republic by force in the 1980s ... had now turned to implementing a soft revolution or a Velvet Revolution, of the type that had transformed authoritarian Communist regimes into nominal liberal democracies" (Kamalipour 2010: 207). The overall assumption is that the Western media has a mission to spoil the Islamic culture of the country from within by imposing Western lifestyle and values. While "clash of civilization discourse fed the fear in the West of Islam, the cultural assault debate nurtured in Iran the fear of the West" (*ibid.*). Westoxicated is a term used to define those individuals who consciously or unconsciously contribute to the spread of Western cultural hegemony.

3 Niavaran Park is vividly presented in the memories of many Iranians in exile and one can easily find a number of references through web pages, blogs, social media groups (two groups under the same names, "Park Niavaran" and "We Are From Niavaran" on Facebook and Instagram) to share their experiences about the park.

4 *Patuq* (gathering place) a term to include gathering places for different purposes entered the Persian social lexicon sometime during the Qajar era (1785–1925) and was originally used to define a space in which the sacred *tough* (a banner used for Muharram celebration rituals) was kept. The placement of the *tough* within a space like mosques, *takaya* (venues for passion plays) or local coffeehouses turned them into a place to attract people to meet and socialize. Although *tough* still is used extensively in Muharram celebrations, the term has expanded to include any type of gathering spots of various natures (Mustawfi 1962: 279; Blookbashi 2007: 372–5; Najafi 2006: 223; Azad Armaki 2006). *Patuqs* not only serve as meeting places for individuals but have also played an important role as an informal institution of networking, dissemination of information and mobilization of masses (professional and otherwise) during various periods of contemporary Iranian history.

5 Insisting on wearing famous brands, which are usually purchased on a visit abroad or smuggled into the country, and make them intentionally visible is regarded as a sign of interest in Western culture, prompted through satellite TV channels and the new media. Considering it as a part of a "Western Cultural Assault," the Iranian government has launched a number of campaigns and implemented a series of unsuccessful efforts (both in policy and practice) to limit the spread of Western dress codes. The phenomenon is probably more visible in northern Tehran (compared to the south part of city) and is interpreted as an expression of "distinction" (in its Bourdieuan sense) and an indication of a "higher" taste.

6 According to Fisher (2008), women tend to get intimacy differently than men do. They get intimacy from face-to-face talking. Women swivel towards each other; they do what we call the "anchoring gaze" and talk. It comes from millions of years of holding a baby in front of her face, cajoling it, reprimanding it, educating it with words. Men tend to get intimacy from side-by-side doing. As soon as one guy looks up, the other guy will look away. I think it comes from millions of years of sitting behind the bush, looking straight ahead, trying to hit that buffalo on the head with a rock. For millions of years, men faced their enemies; they sat side by side with friends. It's deeply embedded in the brain.

7 Security refers to the feeling of being protected against threats and risks that are deliberate and intentional. Safety refers to individual concerns, fears and anxieties in regard to potential violence (accidents that are not intentional), both physical and psychological.

5 How do women perceive women-only parks?

"Safe havens or forbidden zones?"[1]

I was born after the revolution and I've never been without *hijab* outside my home. ... I love to feel the wind blowing in my hair when I'm biking in the park. Such an awesome feeling. I had never experienced it before in my life. I love those parks.

(Mahsa, 21)

Women-only parks might suit some women, but certainly not women like me. As a woman, I want to be visible in society, not pushed into a lavish cage to feel respected and live happily ever after in isolation. I am not a pest to be isolated in order not to infect the rest of society. I am part of society and I want to claim my share. Who granted them the authority to decide how should I live my life?

(Maryam, 27)

The idea of building public parks specifically for women is perceived by many as a token of the dignified status of women, while for others it is part of a larger gender-segregation policy of the Iranian theocratic regime. Hence, those parks as a new phenomenon waked reactions of various groups, from both females and males. When a group of Tehrani inhabitants were asked, some 71 percent "strongly agreed" with the idea of changing some of the gender-mixed parks into women-only ones. There is also a statistically significant difference between men and women, where women are more inclined toward the idea than men (Tehran Municipality Office of Social and Cultural Research 2011).[2] In another study, some 93 percent of a group composed of 614 respondents also "strongly agreed" or "agreed" with the establishment of women-only parks in Tehran (Kawsari 2008). While some groups resented the idea, others perceived it as a meager opportunity for women to extend their presence in the public space. Such variation is also clearly visible throughout data collected and analyzed in this volume. Two main categories emerged, with subsidiary subcategories; each of them will be discussed in detail in this chapter. The first category includes women who are proponents of the idea of building women-only parks, while the second category includes women who resent the concept of building gender-specific spaces such as women-only parks, or find the idea more harmful than beneficial. However, both groups note that they are aware of the need for the presence of women in such public spaces as parks and perceive that to be pivotal. Each group, however, provides different ideas and conditions for such presence.

While there is variation between the groups of proponents in relation to the idea of women-only parks, age and education are among the decisive factors affecting the views toward those parks. Older women are more prone to agree with the notion of women-only parks and they are more contented with the amenities and facilities in the parks compared to the younger generations (Tehran Municipality Office of Social and Cultural Research 2011; Kawsari 2008). One explanation might be that older women are more likely to abide by the rules of *'urf* compared to younger ones. Also, some 48 percent of those who agreed with women-only parks and who are satisfied with such spaces and their services are those with a lower-secondary education (*ibid.*: 48). Hence, one may argue that the level of education can affect one's awareness about rights and choices and willingness to be in gender-mixed spaces rather than segregated ones.

"I love to feel the wind blowing in my hair": proponents of women-only parks

The main argument within the first group of women – proponents of the idea of building women-only parks – considers the building of women-only parks to be a token of respect to women and a legitimate response to their needs and demands, and an endeavor to include and normalize the presence of women in all aspects of public life. The common denominator of this group rests on the premise that they find the idea of a women-only park constructive and positive; and their criticisms are not directed toward gendered spaces *per se*, rather toward the functional failures and practical constraints in the parks. Criticisms target both "not understanding the needs of women for a space and failures in designs, lack of amenities and services for women," and "failing to foster a sense of place" in its users. This includes the sensory experiences as well as perceptions of the place. Given the in-group variations, three main categories emerge within the group of proponents of women-only parks.

The first category includes a group of women, who:

- approve of the idea of building women-only parks;
- are content with the way the parks are designed and constructed; and
- use the parks regularly.

This group is the main users of women-only parks. They visit the parks on a regular basis and are satisfied with the space and amenities provided. Access to a wide range of amenities and services and enjoying the freedom of an open space, otherwise impossible anywhere else, seems to lend a unique experience to this group of users. They repeatedly note the "observance of the Islamic values," a "sense of freedom and comfort," "the beauty and feeling of joy that the green natural environment gives" and the "low cost amenities" of the parks among the advantages which help to "fulfill their needs." Although this group sometimes would identify weaknesses and problems with the parks, they are more prone to see that in line with their religious beliefs and enlarge the strengths and potential of the parks.

A large proportion of this group is composed of women with stronger religious ties and those who support the Islamic government. This group finds the parks to be legitimate institutions in accordance with *sharia* and the ideal role of women in an Islamic society. This is also a group that does not have any problem with the segregation policy of the theocratic government and compulsory *hijab* in the public realm.

The second category encompasses women who:

- agree with the idea of women-only parks;
- however, they are not satisfied with the parks and their amenities and services;
- yet they use the parks on a regular basis.

Within this group, there is a sense of "appreciation toward the efforts of the authorities to accommodate the needs of women to create spaces for them," but there is also a certain level of expectation and criticism toward the parks and their amenities. They are usually more opinionated and outspoken than the first group about additional needs and improvements both in the infrastructure and administration. Their criticisms, however, stem from their own personal experiences and target mostly their own interests and needs. They regularly use the parks but find the "problems and flaws" major hindrances to fully utilizing and enjoying the parks' full capacity.

Since the complaints and criticisms are based on personal experiences, the complaints are often very personalized and are inconsistent or somehow irrelevant to the parks. The criticisms and complaints vary significantly from one person to next and are different in different parks and cities. Taking from the responses, the criticisms include both complaints about the flaws and weaknesses, and suggestions for improvements in the amenities, services and regulations. One of the most frequent complaints is the low accessibility and inappropriate location of the parks and placement of the facilities and services inside them, which seems to be a predominant issue for many users. The problem seems especially pertinent in larger cities. Low accessibility could arise due to various factors including poor connectivity – as a result of dysfunction or lack of public transportation networks across the city – and decrease the number of visitors going to the parks. The low intensity of the crowd in a public space affects its vitality, a problem which is noted by many users in women-only parks. They "do not feel safe" in the places that are not connected to the surrounding fabric or do not have easy and rapid access to the public transportation network of the city. A young woman who is 26 years old raises her concern about the neighborhood where the Pardis Banvan Park in Tehran is located and explains her solution to the problem: she "needs to ask her husband each time to give her ride to the park." This is inconsistent with the original idea of building women-only parks as spaces to promote women's visibility and presence in public and to encourage them to function independently.

Since women-only parks must meet the criteria of enclosure in order not be visible from the outside, they are usually located far from city centers or are built on the outskirts of towns. The geographical location of the parks has strongly affected the accessibility through a longer commuting time for many users. The commuting time is even longer in larger cities with crowded streets and heavy traffic. Many women point out the long distances between their homes and the parks and the longer commuting time as the main reason that prevents them from using the parks more frequently. A woman in Bihisht Mardaran Park in Tehran notes,

> The park is too far from my place. With such limited operating hours, I rarely get the chance to use it. Doesn't happen more than two or three times a year. I wish I had one of these parks closer to where I live.

Many women in this subgroup note that the number of women-only parks is not enough to satisfy the need and that the authorities should think of building one in each district or neighborhood. In this way, they argue, "the women-only park of each district would easily be accessible to the residents of the same neighborhood." A woman in Bihisht Madaran Park argues,

> It is kind of a problem that there are just a few [women-only] parks in the entire city. They [the authorities] must build more parks in different parts of the city so that everyone can have one in their neighborhood. Many women are not able to use them regularly, simply because there is not one close to where they live.

In addition to low connectivity, a lack of functioning public transportation fabric and longer commuting times, incompatible functional characteristics of the surroundings areas are also among the factors affecting the patterns of park usage among women. For instance, users at Pardis Banvan Park in Tehran argue that despite the fact that variation of amenities and services in the park is attractive, the location of the park and the problems that may entail overrides all those advantages. When women consider the surrounding areas unsafe, they try to avoid either being in that neighborhood or they walk as a group. The negative reputation of the area along with a lack of public activities or compatible functions in the neighborhood discourage the use of the parks by women, especially those commuting from farther parts of the city and those who are not familiar with the neighborhood. One woman says,

> It would be nice if each municipality built a park in its district. These parks are too far away and there are too few of them. The long commutes make us dependent on private cars or someone to give us a lift.

Lack of amenities and services is also among the main concerns and complaints of many users at women-only parks. Smaller parks are more likely to suffer from the lack of amenities, partially due to the scarcity of space to accommodate them. Pardis Park in Isfahan and Park Banvan in Rasht are instances of such small women-only parks which lack the amenities to respond to the needs of their users. A 31-year-old woman who was visiting Pardis Park in Isfahan for the first time reflects, "You call this a park? Looks like a large room without a roof." She continues, "it's poor in all aspects. ... More like a garden whose design has gone terribly wrong." Departing from the users' perspective, while the lack of or limited amenities as well as overcrowding in smaller parks are more visible problems, the larger parks suffer from uneven distribution and inappropriate location of amenities. These are repeatedly referred to as being among the main hindrances in using women-only parks. Limited, dispersed, ill-located and inaccessible restrooms, hygiene facilities, food stands and cafés inside the parks in Tehran are also among the main complaints which are repeatedly noted by women. A young woman at Bihisht Madaran says,

> There are only very few spots for drinking water in this park and they are usually far from those parts of the park visited and used most. There is only a single, tiny shop, which doesn't carry most of the basic supplies one may need.

Another woman reflects on the uneven distribution of the basic facilities at Bihisht Madaran:

> Amenities and furniture are not distributed evenly and equally in different parts of the park. Some parts are over-furnished while there is hardly any furniture in other parts, which has resulted in over-crowdedness in some parts at the expense of under-usage of others.

The physical structure and ill placement of the amenities and services are not the only concerns noticed and noted by users. Complaints and recommendations regarding surveillance systems and regulations at women-only parks are repeatedly brought up by users. As previously noted, every park has adopted its own distinct strategy to manage and control the interior space of the park. Some parks invested in hard surveillance systems – checkpoints including body searches and inspection of belongings at the very entrance of the parks in order to be granted permission to enter the park; others employ soft control methods reminding women about the prohibited items and forbidden activities. Using hard control systems has triggered discontent and advanced the dissatisfaction of users at the parks. Many women consider them disrespectful and complain about the attitudes of security guards and disproportionate surveillance systems. They argue that such methods complicate usage of the parks and create feelings of uneasiness and "a sense of being watched." Hence, "an easy and straightforward act of entering the park to enjoy a moment of tranquility and pleasure would resemble a complicated process of entering a military zone." A 29-year-old woman points out that the surveillance and control methods are a serious obstacle:

To me the presence of so many security officers at the very entrance is one of the hindrances to using the park. It makes me uncomfortable. It is more like entering a jail. Or, perhaps more like a prisoner being allowed to take a short break in the open air.

While there is an awareness about the responsibility of the guards to maintain the safety of users and security in the parks – which leads to a certain level of empathy with the guards – women are also constantly complaining about the way they are treated. For the majority of women, the hard surveillance methods at the entrance are disproportional, disrespectful, inappropriate and exhausting. They particularly note long queues when waiting for the search process, which are troublesome and unnecessary. A young woman reflects on her unpleasant experience of entering Pardis Banvan as follows:

I totally understand that it's part of the regulation and they have to check everybody's belongings. But just look at the narrow pathway at the entrance, which forces women to stand in a single line, I found it annoying and insulting. They are treating us like cattle.

The age limit for young boys accompanying their mothers to women-only parks is another recurrent topic of complaint and a serious obstacle to using the parks, especially for young mothers. According to the regulations of women-only parks, boys older than five years are not allowed to enter the parks. Increasing the age limit for boys accompanying their mothers into the parks is among the main demands of many users at various women-only parks. Many argue that the regulation stands in conflict with the religious tenet on the so-called "age of distinction" (*sinn-i tamiz*). The creed suggests that boys under the age of nine are not in the position to distinguish their sexual desire and their presence among women does not require observing dress code or practice of *hijab*. They are allowed to enjoy mingling freely with the opposite sex or to be in an all-woman environment. A young mother at Bihisht Madaran reflects,

I can't bring my six-year-old son to the park with me. It's prohibited if a boy is older than five. I'd like to ask them if they really think a six-year-old boy understands anything about women's body. They seem more catholic than the pope. They even ban what *sharia* allows.

Many women argue that the parks are not child-friendly and mothers with young children are not able to enjoy the parks in the company of their infants. They always have to stay alert and not let children out of sight, since the parks are not safe, and they are not designed to accommodate the needs of young children. A young mother notes,

They have built a park for women, yet nobody could think that one day a mother with a baby may come to the park and she may need to change her baby. There is not a changing room or a private corner that you can use to change your baby.

In general, comments of this group include complaints, reflections, recommendations for improvements and suggestions mostly based on their own personal experiences of various complications in using the parks.

The third category, however, includes women, who:

* approve of the idea of women-only parks and find those parks useful additions to encourage women out of the home;
* however, despite their will, they have not been able to visit or use any of them and their ideas are not based on first-hand experience of the parks.

A 40-year-old Tehrani woman describes her first glimpse of Pardis Banvan women-only park's exterior:

I took an airport cab home the other day. The driver was driving on a highway somewhere in southern Tehran, passing by a huge garrison-like installation with tall brick walls in the middle of the freeway junctions. I thought it was a military base or something. But the driver told me that it was a women's park. I had never heard of it. Must be interesting to see how it looks from inside.

A 43-year-old Tehrani woman blames a lack of information about the parks and their services and amenities and notes, "it was a few days ago and only by sheer chance that I learned about special parks for women in Tehran." Other women state that they had already heard about such parks but they did not have any information about their locations and/or operating hours. In addition to the lack of information and ineffective dissemination methods informing women about the parks and introducing their amenities and services, many women in this category give inappropriate locations, limited operating hours of the parks and ignoring the needs of workingwomen as some of the main reasons preventing them from visiting the parks. Since the parks have limited operating hours (operating hours vary in different parks but none of them are open later than 19:30), many, especially workingwomen, rarely have any chance to visit and use the parks. Women's reflections show a recurrent pattern of complaints about closing time in the evenings, especially during long summer days. Many think that the parks are not utilizable during the hot hours of the day and they prefer to visit the parks later in the afternoon or early evening when it is cooler.

Whereas all three categories of women discussed here include proponents of the idea of women-only parks, who consider such parks a response to the needs of Iranian women and a token of respect to women, they provide different reasons that affect their pattern of using the parks.

"I'd never set my foot there": opponents of women-only parks

As noted earlier, women-only parks were turned into a scene of social and political contestation. The opponents of women-only parks are against the idea of building gender-specific spaces in general and perceive the parks as part of a larger gender-segregation policy of the Iranian post-revolutionary theocratic regime. Women who are against building women-only parks (or any other gender-specific spaces) consider them another step toward further separating the sexes in such a highly-segregated society as Iran. Contrary to the previous group who consider the parks a token of respect to women, many women in this category find them socio-politi-cally-engineered institutions in line with a series of policies and actions to restrict women's presence in public. These women express their ideas and feelings about women-only parks in a number of different ways. A 53-year-old opponent reacts to the women-only parks in following way:

> I don't like the idea at all. It's a way to curb women. I always think that women and men are different yet complement each other. They both need to be together to enjoy life and make society function. To me, separating women is a kind of keeping them away from the society. It doesn't give a good feel-ing. It makes me feel isolated. I don't get it. Why should we be afraid of each other? Why should we feel disturbed by the presence of men in the parks? Why should men be excluded so that women can enjoy?

Variation in comments reflects differences in reasoning among various groups of women and reveals the factors which contribute to the discontent of women and ultimately affect their active decision to refrain from using the parks. From an urban design perspective, the notion of the publicness of any given urban space includes its accessibility to all populations. Hence, from a theoretical standpoint, women-only parks violate a fundamental principle of publicness by excluding men from accessing and using public parks. A 50-year-old woman living in Tehran expresses her feelings:

> The current situation of Iranian society is far beyond my grasp. I have a hard time understanding it. This segregation is not acceptable. The time that I grew up was very different. To me the idea of women-only parks is really stupid because the parks are public spaces for everyone. How can they build a park just for women? If they could improve the ordinary parks for women to feel safe and not be afraid of harassment, there would be no need to build women-only parks.

A 30-year-old female architect describes her impression after visiting Bihisht Madaran:

> To me, those parks are superficial and useless ways to solve the problem of women's presence in public space. Women-only parks will create a problem

for women's presence in integrated and mixed spaces. It seems that the government and the municipality as its administrative body aim to intentionally increase the gender gap in society by creating more segregated spaces. This will create an unhealthy society.

Most women in this group are against the idea of women-only parks, yet some use those parks. They argue that while they are not fond of gender-segregated spaces, such places are more convenient compared to the restrictions and difficulties they face in mixed spaces. The restrictions of Iranian women in public along with a series of disciplinary measures which limit the mobility and interactions of women is pointed out repeatedly as one of the main reasons to attract women – despite their will – to those parks. In other words, in the absence of better alternatives for women's presence in public spaces and a lack of social networks and activities, women-only parks seem to present themselves as the sole option. Lack of safety and security, sexual harassment – both verbal and physical – and the "uncanny feeling of being gazed at" while doing exercises in mixed parks are among the reasons encouraging women to use women-only parks. A young, 26-year-old, physician in Tehran reflects on the lack of safety in mixed parks, which discourages her from using them:

> I'm totally against the idea of building parks only for women. I would prefer it if they would improve safety and security within the mixed ones. This way, people would feel safe and comfortable in the parks in their neighborhoods and use them more often rather than commuting long hours to use a women-only park at the other end of the city. Until then, I have no choice but to use women-only parks.

Despite the fact that all women in this category resent the notion of building women-only parks, they choose different strategies of whether or not to use them. Hence, the "opponent group" includes several categories based on their social and political stands, their preferences to use the parks, the extent of satisfaction from the parks and their respective amenities and facilities.

The first category includes women who:

- resent the idea of building women-only parks;
- but consider it to be a possibility in such an otherwise male-dominated society as Iran.

Despite their resentment toward building any gender-specific spaces, this group of women uses the parks and in general is satisfied with the quality of the built spaces and provided services. For many of them, the physical quality, design, amenities and services provided are at an acceptable level and the only concern is that of social factors. They adamantly argue against the negative social impacts of gender segregation in society. A young woman blames the "domination of normative values and their restrictive nature" as a main reason encouraging women to use separate parks of their own. She explains her feelings about women-only parks:

> as someone who doesn't practice any religion, I'm totally against the idea of women-only parks. But as a woman living in an Islamic country, with all those limitations and restrictions, it doesn't seem like a bad idea.

Another woman considers women-only parks a window of possibility for women to make their presence in public possible and expand their social interactions. Hence, women-only parks function as public institutions to promote social interactions and provide possibilities for the public presence of women – albeit in a homosocial environment. The networks built in such environment, however, usually extend beyond those spaces and continue to expand over time. A woman reflects on her regular visits to parks:

> I totally disagree with the idea of separating women and men in society. But in the current situation in Iran these parks open up a decent possibility to women. It's a nice place if women want to hold gatherings with their extended family and friends; otherwise they would have to meet at home. They don't have any other option, though. In the regular parks you can see older women more often but in women-only parks, younger women are more visible.

Regardless of the views on women-only parks, amenities, activities, services and the uniqueness of experience at the parks are among the factors that contribute to women's willingness to visit and use women-only parks. Doing exercises in a green open space and spending time with friends in a safe and comfortable environment are instances of such activities, which Iranian women do not have the possibility of experiencing in other public spaces. Moreover, many women in this category agree that such parks provide ample possibilities for women with stronger religious ties or those who are living in traditional families. A young woman explains that despite her own negative view toward women-only parks, she respects the rights and wishes of other women who "prefer to use such segregated spaces." She further argues that, such parks may not satisfy her appetite for the social presence of women in public, but it might be the only possibility for other groups of women, especially those with religious bonds, to enjoy public spaces and their respective services. She states,

> I think it's very important to listen to opinions of different groups of women with various cultural and religious backgrounds. I just saw some women take off their *chadur* but continue wearing *hijab* inside the park. I don't have the liberty to tell others to enjoy their freedom the way I like. You'll see variations among women in their views toward the parks and the way they use them, and it is because of the differences in their backgrounds and the ways they think.

Another woman expresses her opinion:

> On the one hand, I found it part of the government's Islamization program through which they want to separate women and men everywhere in the entire society. I strongly resent the idea. But on the other hand, I can see that these parks have created safe and comfortable spaces for women to enjoy their leisure time. It's not fair to turn a blind eye on that and pretend that they don't exist.

Women also show their discontent about the quality of parks and services. The overall quality of parks varies from one park to the next, depending on the size and location, which determines satisfaction among users. While reflecting on the advantages of a homosocial environment, they also reflect critically on the weaknesses and flaws. A young woman addresses a few of such problems in relation to Bihisht Madaran Park in Tehran:

> The only good thing about a women-only park is the freedom to dress inside the park. But this is not enough to turn this park into my favorite place … I usually use the park to exercise. There is a training hall but there are no showers, lockers, restrooms and food or drink venders. A lack of basic services or ill placement of them, which requires one to take a long walk to find and use them, is one of the main problems with the park.

The second category within the opponent group includes women who:

- resent the idea of women-only spaces;
- make a conscious and active decision not to use women-only parks.

This group argues that women-only parks broaden the already wide gap between the sexes in society. They note that in the absence of possibilities for a natural interaction between women and men and with the implementation of systematic gender-specific spaces from pools to schools, universities and taxis, women-only parks add another layer to further separate them. This exacerbates the problem of women functioning under normal circumstances and gradually marginalizes women in public space. Many argue that the efforts to marginalize women are part of a larger policy to push women into the private sphere of homes and to encourage them to return to their roles as mothers and wives. A young woman reflects,

> With all respect to those who practice religion and may favor those parks, I think they will increase the gap between women and men in society. I don't consider it a good initiative for Iranian society and culture. In the long run it will create more harm than good. Women and men should learn to live together and respect each other. How can they learn this when they don't see each other?

Some women, including a group of women's rights activists and public intellectuals, openly criticize the idea of women-only parks. Zahra Minooie a women's rights activist perceives it as the extension of the gender-segregation policy in Iran and argues,

> taking it from a macro perspective and with reference to the gender-segregation policy of the government which is already implemented in other sectors such as education and transportation, we cannot be optimistic about the outcomes and impacts of those parks. This is more of a humiliation of women than a respect of them, as the authorities try to make us believe.

Another young woman raises a concern which is shared by many others:

> To me, forbidding men to enter such parks and encouraging women to use them more and more rather than the mixed parks will affect the social skills of young women in interacting with men. Imagine a girl who rides on the women section of a bus or metro car to school, spends all her time, from morning to afternoon, day in day out, in an all-women environment and in her leisure time goes to a women-only park in the company of her mother, playing and spending time with other girls. Where is this girl supposed to learn to interact with boys? There is no such possibility to learn to interact with males in this society. Interacting with male members of the family and relatives within the privacy of the home environment does not replace interaction in society. ... Social interaction and the ability to function in the public sphere are very important and must be taught through a healthy relationship between girls and boys at a young age. This segregation will not help girls to function normally in society when they grow into adults.

A female architect and urban planner who actively distances herself from using the women-only parks explains her reasons:

> The normal social dynamic, which makes every society and any given public space vibrant and dynamic, is lacking in these parks. It looks much like an artificial environment created for a specific purpose, not necessarily for the well-being of women. Despite the merits claimed by the authorities who created those parks and the women who believed them, I am not convinced that a closed, segregated, fortress-like space filled with women of a certain age and background can contribute positively to a healthy society. A healthy public space is a microcosm of the society including people of all backgrounds, female and male, who interact naturally and respectfully. It looks like they have extended the home environment in which women can sit and continue their girl talk. I'd never set my foot there.

Many of the women who are against the notion of women-only parks also argue against the project of forced Islamization, which entails compulsory *hijab* and gender-segregated public spaces. This group perceives the women-only parks as institutions to frame, formalize and internalize mass Islamization. Many take a rights-based perspective and argue based on one's basic human rights. For them the gendered spaces are the rendition of the patriarchal normative that undermines and minimizes the role of women in society. Hence, gendered spaces, such as parks, are perceived as "metropolitan *andaruni*" designated to women. For them, women-only parks are the extensions of the private sphere. Women are regarded as the collective *namus* to be guarded in order to guarantee the moral well-being of society.

Notes

1 Ziba (54 years old).
2 The research used the evaluation method and was commissioned and conducted by Tehran Municipality's Office of Social and Cultural Research. The sample of 1,161 Tehrani respondents (57 percent women and 44 percent men) from age group 18 and above was randomly selected from all 22 districts in Tehran. While the study claims to be representative, the sample seems skewed. It is composed of 41 percent housewives and 12.5 retired people, and some 72 percent "with secondary education or lower." Since the findings of these studies are used to justify building more women-only parks, their findings should be used with a certain degree of caution and reservation.

6 Public urban space, the female body and segregation

A conclusion

> Body-reflexive practices ... are not internal to the individual. They involve social relations and symbolism; they may well involve large-scale institutions. Particular versions of masculinity [and femininity] are constituted in their circuits as meaningful bodies and embodied meanings. Through body-reflexive practices, more than individual lives are formed: a social world is formed.
>
> *(Connell 2005: 64)*

The female body is singled out as the core of social engineering and the manifestation of political ideology in any normative-driven society. The social management of the body (and all respective issues related to various forms of relationships and interactions) serves as the foci of urban experience – an interplay of social relations, urban forms and subjective positions within an urban context. Also, spatial arrangements enhance the reproduction and structures of gender and sexual relationships, and the articulation of identities. Socially normative lines of division and dominance of values are shaped and reinforced by urban environments, and individuals find spaces in city venues to perform or express such identities. To think about gender and sexuality in the city, then, is to think about the interaction of spatial practice, social difference and symbolic associations in urban contexts. Setting up gender and sexuality in the urban context is partly a question of putting bodies in space, which also indicates how embodied subjects are located within more general social structures and relationships. Gender and sexuality, after all, are not defined by the limits of the individual body, they involve social relations that extend across and are shaped by space (Tonkiss 2005: 92–4). Hence, the presence of the female body in the public domain and in the context of a heavily normative-laden society, such as the post-revolutionary Iran, could be contextualized through what Najmabadi (1991) calls "ideologisation and instrumentalization of the woman question" wherein the female body becomes the symbolic location of normative values and cultural practices.

In her study on the so-called "pleasure garden" model of urban parks in the early twentieth century United States – a phenomenon similar to women-only parks – Cranz (1980: 79) notes that, "women as a category have not been perceived as an urban problem, park policymakers have used females primarily to help ameliorate other problems which disrupted social order." This, as much as it was related to the invisibility of women in public, was also the result of a series of civil society's achievements on gender equality, as it opened up urban public space for women. In Iran, on the other hand, what Najmabadi (1991: 47–8) labels the "woman question" (*mas'ali-yi zan*) meaning the new problematic place of women in a modern society, which has been shaped as a central part of an emerging climate of political ideas and social concerns in a desperate call for the creation, re-appropriation and redefinition of a new Islamic political alternative. The Iranian Islamic government rested the argument on the expectation that solving the "problem of" the presence of women in public would contribute to the solution of other societal problems.

For the Iranian theocratic regime, parks were not considered significantly contested public spaces. However, the significance of parks as primary public spaces and ideological battlegrounds in the process of the enforcement of the compulsory *hijab* law became obvious when briefly, after the revolution, the newly founded Bureau for Combating Moral Corruption "stepped up its activities in response to 'popular demand' for *hijab* by closing down a number of parks in Tehran for lack of observance of Islamic code of conduct by some women." It warned the "sisters who intend to use the parks to observe Islamic *hijab* or be prevented from entering the parks" (Kayhan 1981). Like Cranz's (1980: 80) argument about the debate over public parks in the United States where the "women's central role in the family meant that urban reform and women were often linked explicitly," the solution of Iranian women's presence in public was strictly tied to the image of the ideal woman in a Muslim society. As detailed in this volume, ideological reasons neither entreated the return of women into public parks nor provoked a solution to mediate between religious mandates and a modern lifestyle. Despite socio-cultural differences, however, the similarities between gender and the public parks debate in the United States in the early twentieth century and post-revolutionary Iran is striking. Just as "the park commissioners [in the United States], not wanting to compete with the home as the proper mechanism of moral reform, underscored the important ways in which the park would help reinforce the family unit" (Women News and Analysis Agency 2013), Iranians emphasized the role of healthy mothers in reproducing a generation of Muslims. In Cranz's (1980: 79–95) words, "if the home [is] the fortress of morality, why should women be brought into the public sphere at all?" The answer is that "women would set a tone which would demand high standards by anyone. ... No laws and no police force will do it. A park would educate the family into the ways of disease and vice" (Will 1894: 277). While the setting for women in public was convincingly safe, a respectable woman "was not supposed to go out alone. ... and the parks must be made a safe resort for unprotected women" (Low *et al.* 2005: 81). Hence, gender segregation, gender-role stereotyping and unequal treatment are closely linked to the point of discrimination. The theory of separating women and men presumes that they are different and at the same time guarantees that treatment of the sexes will not be equal.

However, the gender-segregation policies of the Iranian theocratic administration were not limited to the segregation of physical spaces; they also extended to an array of social settings including women's physical activities and sports.

With regard to sport, segregation led to the almost complete cessation of women's participation for almost two decades. Paradoxically, the very policy that banned outdoor activities for women in public for more than two decades became a pretext for building women-only parks in Iran. The ideological tenets of this segregation policy, which lead to a denial of resources for women's physical activities, emerge vividly in this statement from the then head of the Physical Education Organization:

> In the Islamic Republic, women don't have time for sport. Women's main responsibility is housekeeping and child rearing and they can get a lot of exercise from doing these. The circumstances in our society at the moment do not allow women to spend their time on sport or on campaigning for it. The country has other priorities than spending money on women's sport facilities.
> (Kayhan Havaie 1979 as cited in Paidar 1995: 341).

Despite changes in both the policies and practices during the last 37 years following the revolution, this has remained intact as the dominant discourse.

> [T]he issue of women is not separable from the issue of family. ... One of the greatest mistakes of western thoughts about the issues of women is this sexual equality. Justice is a legitimate concept, but equality is sometimes legitimate and sometimes illegitimate. Why should we separate an individual who has been built for a particular domain – in terms of one's natural make up, whether physical or emotional – from that particular domain and drag her towards another domain which Allah the Exalted has built for another make-up? Why should we do this? What reason do we have for doing so? What kind of sympathy is this? Why should women be entrusted with carrying out male tasks? What kind of honor is to have women carry out male tasks?
> (Khamenie 2014)

However, limiting the presence of women in public meant forging a new space through the redefinition of women's traditional roles in the public domain. Women's presence in public was welcomed as long as it did not disturb the society's traditional patriarchal structure, and as long as it did not pose a threat to traditional masculinity, and did not introduce a new definition of the femininity, especially one affected by Western feminist discourse. Hence, women's presence in public, except on the occasions commensurate with the image constructed by Iranian new-traditionalists, was perceived as undesirable and "was blamed for the weakening of the traditional institution of family and believed to challenge the traditional role of man as the provider, for which they should have the priority for employment and hence entitled to a higher salary" (Najmabadi 1991: 67). The post-revolutionary social engineering, which aimed for Islamization of the entire society, resulted in a gender-based division of labor in the family and society, in which both women and men preferred their "natural" officially defined roles. Women were exalted as mothers of the nation, whose honor was perceived a collective *namus* to be guarded both in private and in public. Women, thus, were considered the nurturers of the Islamic nation based on their role as creators of the family.

In many parts of the world "mother as nation" naturalized the ideology of motherhood – in forcing women to "want" to be officially defined mothers (heterosexual, biological, stay-at-home). This restricted mothers to the private realm, rather than allowing them a public voice (Bernstein 2008). In Iran, "mother as nation" was a nurtured construction of an Islamic discourse based on power dynamics between national Shiite identity and the global *ummah*. Hence, the merging of mother imagery of the nation-state on the one hand, and the idealized Muslim woman on the other, reflected what Ranchod-Nilsson and Tetreault (2000: 16) call, "women's corporal and cultural roles in reproducing the next generation." Thus the "politics of reproduction" is a force synthesizing these two perspectives – "the local and the global" with multiple levels on which reproductive practices, policies, and politics so often depend (Ginsburg and Rapp 1991). It is similar to the socialist notion of the "heroine mother" (Anton 2008) where the construction of a national maternal identity became the state's concern and led to a comprehensive public discourse centered around women's national duty to reproduce, disseminated through educational institutions, official television channels, films, radio shows and literature to exclude all other counter-narratives. However, the image of moral mothers and the reflection of motherhood as a national discourse were used to redefine and reshape the ideal female image and transform it into a national and collective concern.

The indirect pressure on women to stay at home and become better mothers, despite achievements in higher education and ample opportunities in the labor market, offered a new challenge for Iranian women. Paradoxically, the policy which assumed a new public role for women by providing them access to higher education also encouraged them to stay home, limited their access to public spaces and defined women as least in need of spaces for physical and leisure activities. The policy that once ignored the fact that women were actually the group who needed such activities, after two decades, argued and justified the establishment of spaces for women's physical activities. Multiple reports by the Iranian medical corps and a large body of scientific research, warned the public about the "tsunami of epidermis diseases" attributed to vitamin D deficiency resulting from lack of sunlight exposure (mainly due to the compulsory *hijab*) and lack of physical activity (a result of limiting women's outdoor presence). This paved the way for the construction of parallel gendered spaces in Iran.

In practice, however, the Iranian gendered space policy was in many ways similar to "the colonial project ... [which] was about seizing control of geographical areas to produce new spatial relations of boundaries and hierarchies ... extended to the classification of people into categories ... and manufacturing of a reservoir of cultural imaginaries" (Mbembe 2003). In the same vein, the Iranian post-revolutionary political regime used gendered spatially regimented division as part of a larger Islamization project.

Urban studies literature on the notion of the public–private spatial dichotomy suggest that, using the criteria of access, agency and interest, a space can be considered public if it is controlled by the public authorities, concerns the people as a whole, is open or available to them and is used or shared by all the members of a community. Urban, open public spaces, therefore, have usually been defined as multi-purpose spaces distinguishable from, and mediating between, the demarcated territories of households and individuals. One may argue, however, that a definition that departs from such a dichotomy imposes limitations and ambiguities on the notion of public–private in a wider array of social contexts. A more inclusive definition must be based on relations between the two rather than on the legitimacy of power, agency or the type of activities in each. Madanipour (2003) suggests examining a space beyond its physical characteristics to include social interaction and socio-psychological space. Therefore, public space is defined in relation to the private, which necessitates an understanding of space that incorporates the nature of social interaction within it, rather than the physical division.

> Depending on the way the private realm is defined (mind, body, property, home), the public sphere finds a related but opposite meaning. If mind is the private realm, the outside world is the public. If the body is the private realm, the other bodies constitute the public. If private property is the private realm, what lies outside private possession and control is the public. If the household is the private realm, the larger organizations and the rest of society is the public. The private realm can be one or a number of these layers and as such the public realm can be formed of a number of such layers.
>
> (Madanipour 2003: 98–9)

However, the gendered nature of such public spaces as women-only parks challenges such elaboration of the public–private sphere, as it fails to account for the dichotomy analyzed in this volume. The theocratic political system in Iran drives the ambiguity even further and complicates the public–private division even more. The political theory based on *vilayat-i faqih* (authority of the jurisconsult) stipulates the *faqih* as the legitimate political figure who gains his power from a divine source, and hence is not subject to lay questioning. The practical purview of this theory includes the unclear border between public and private domains. Despite the fundamental difference between the nature of divine and human powers, the *faqih* extends the divine lordship (power of God over man) through the doctrine of acquisition (*kasb*),[1] rendering it to human power (power of man over man). Consequently, the private domain becomes the extension of public (and vice versa) and hence the legitimate domain of the *faqih's* power. By extension, any space (whether private or public) becomes a domain of legal scrutiny by the Islamic state. With the blurred border between public and private suggested by the theory of *vilayat-i faqih* and the all-encompassing power of *faqih* over laymen, the publicness of the space is defined

within the broad frameworks of state and society. A public space is therefore often provided and managed by the state and is used by the society as a
whole. ... [It] may or may not legitimately represent or serve a community
... may or may not be willing or able to use a particular space for functional,
symbolic or any other reasons ... places outside the boundaries of individual
or small group control, mediating between private spaces and used for a variety of often overlapping functional and symbolic purposes ... distinguishable
from, and mediating between, the demarcated territories of households and
individuals.

(Madanipour 2003: 98–9)

The notion of '*urf* as unwritten socio-moral contracts practiced widely in the
Muslim world further complicates the matter.

Hence, the imposition of *hijab* and the gender-segregation policy were part
of a larger practice of biopolitics. The Islamic state aimed for total control of all
female–male relationships in public, and to a larger extent in private. In Stoler's
(2001) words, "matters of intimacy become the matters of state" as part of the
process of governmentality. The totalitarian nature of *vilayat-i faqih* extended
throughout society, both public and private, including not only biopolitics, but
also necropolitics through the Islamic doctrine of *qisas* (legal retribution) and the
penal code adopted from it.

Despite the fact that *hijab* and gender segregation were informally institutionalized and with the tacit rule of '*urf* practiced on a voluntary basis across various
segments of Iranian society for centuries, the Islamic state subjected it to government intrusion – "as the family was turned into a political institution, the violation
of its sanctity became a political crime. The importance of Islamic gender relation
to the establishment of Islamic society required heavy punishment for its offenders" (Paidar 1995) – and made it subject to *qisas*. The new term "*bad-hijab*" was
coined to criminalize the inadequate coverage of a woman's body and the failure
to comply with the Islamic dress code and its main symbol: the *chadur*.

Thus, gendering urban space to achieve and enhance social Islamization suggests "the apartheid city" which, as Hansen notes, "perfected colonial forms of
governance by converting race to space" (Hansen 2008: 101). In a similar vein,
the Iranian regime of power enhanced Islamization through rendering gender into
space. The body becomes the field[2] of governmentality, as it is:

directly involved in a political field; power relations have an immediate hold
upon it; they invest it, mark it, train it, torture it, force it to carry out tasks, to
perform ceremonies, to emit signs.

[...]

the body becomes a useful force ... may be calculated, organized, technically
thought out; it may be subtle, make use neither of weapons nor of terror and
yet remain of a physical order.

(Foucault 1977: 26)

The women-only parks nurtured debates and discussions about women in the public realm in Iran and the extent of institutions and practices for gender segregation in society. Critiques about the women-only parks are not solely made in reference to the impact of those parks on the social skills of women or the effects of segregation policy. Some scholars refer to the contradictions in the process of theorizing the need and the implementation of women-only parks and perceive the arguments by the authorities far from trustworthy. For instance, they note that, while the needs of women is argued to be the main principal behind the idea of women-only parks, no woman was consulted or engaged in the process of design or planning. They argue that the idea, like the rest of the segregation policy, is a patriarchal practice where men decide women's needs on their behalf. The same criticism is also made in reference to the shorter operating hours of the women-only parks, the reason for which was to use male manpower for gardening, watering and other maintenance. "Do they really need to close parks earlier so that *men* can water the plants in the parks," argues a young woman. Or as another woman notes,

> Using men for such jobs shows that the authorities: a) believe that these type of jobs are not appropriate for women – they are man jobs; and, b) they still think that men are the breadwinners for the family and they should be prioritized over women. And they want us to believe that they care about women and their needs?

Women-only parks sparked a heated debate across various groups in Iran and the proponents and opponents of the parks are not limited to women. Many male activists and scholars have equally engaged in the debate over the women-only parks. For instance, Qarayi Moqaddam, a male sociologist, argues that women-only parks – through providing spaces for removing *hijab* – function as venues for social catharsis among women and must be welcomed and appreciated to assure "the health of the society" (Ganji 2011). Ali Entezari a pro-government male sociologist, nevertheless, criticizes Moqaddam and maintains that,

> we must move in a direction to institutionalize *hijab* across the entire society. Women must be encouraged to observe *hijab* everywhere and under all circumstances in public. By providing space to experience removing their *hijab*, the parks create confusion among women and interfere in implementing the overarching objective of women observing *hijab* in public at all times. They serve more publicity and propaganda purposes. They harm more than doing any good.
>
> (*ibid.*)

Entrance of men into the debate on women-only parks revealed new dimensions within the question of women in public space. A female sociologist (43 years old), however, reflects on Entezari's idea:

It's dangerous to think that way. This means that we [the authorities] should not let women experience freedom, because it will be difficult to convince them to stay confined. They know that the project of compulsory *hijab* and forced Islamization is doomed to failure and the only way to maintain it is not to let women to know how freedom feels. He [Entezari] should not worry about the impact of parks in giving a glimpse of freedom to women. Sense of freedom cannot be fully experienced in captivity. The real feeling kicks in when a woman removes her *hijab* in the presence of men, not in a glassed space and in the presence of some other fellow women. It is a manipulated and a fake experience, far from real. ... They keep telling us that the women-only environments are a new trend in Western societies and, across the United States and Europe, women demand more women-only parks. I am not in a position to verify this information. But what they don't tell us is that an American or a European woman can appear in any park she wishes, dress the way she likes and do whatever activity she pleases. They have choices and they choose what they wish. We don't have any choice. They tell us: we have decided for you that you'll feel free this way, within this environment. Take it or leave it.

Another young female sociologist (28 years old) reflects on the parks:

A women-only environment – whether a park or any other space – doesn't attract me at all. A space composed of both women and men looks more real and appealing to me. I don't understand what would be the attraction of a caged space where women are expected to interact with each other and be happy where men cannot see them. Such a thing is not a solution. The real solution is to invest in and promote the culture of respect and co-existence so that women feel safe in the mixed parks. To respect my needs as a woman does not mean that municipality isolates me from men and keeps me somewhere so that no man can reach me. I personally – and many of my generation share this with me – feel quite comfortable in a mixed environment. Authorities make this false assumption that we are not comfortable and assume themselves our custodians to decide for us that we will feel better if separated. Society would look far better if they could change their mentality.

In explaining why the female body in public becomes a field of contestation between women and men, one may refer to Jorun Solheim's notion of "the open body." Solheim (2001) argues that in reading the semiotics and symbolism of the female body in the context of heterosexual normalcy, the female body is "symbolically coded as open." While Solheim builds her theory around the notion of the physical openness of the female body, here it is more extended into the domain of the social and normative. Hence, the social situation and presence of women in public in such societies with patriarchal structures becomes the area of *namus* (honor) – both collective and individual – and decision/policy making. While the Solhemian notion of "world of experience" is physical, it is more social in the context of the present volume. The openness of the female body makes it a subject of honor for family and kin and the core of the management of desire for the state, and is legitimized with reference to religion and culture. To quote Solheim:

> the physical "world of experience" as something in itself is mediated through a set of symbolic beliefs ... and this "symbolic order" is based on a special kind of meaning construction, which seems to revolve around a specific representation of the female.
>
> (Solheim 2001: 12)

Giti Etemad (Aramesh 2012) notes that gender-specific public spaces are pre-modern institutions, which ignore the fact that women are part of the public domain. She argues that such spaces create hindrances for normal interaction between women and men and force them to find new ways to meet up. For instance, they use the parks in the northern part of the city (Tehran) where the environment is more tolerant. Or they even utilize religious rituals like passion plays – popular practices among *Shi'a* Muslims – or religious processions of Muharram as dating opportunities. Azar Tashakor (*ibid.*) argues that the government exploits a feminist discourse on the rights of women to public spaces to institutionalize women-only spaces. In reality, however, it creates gender-based ghettos. Limiting women's access to physical public space has encouraged them to benefit from the advantages of virtual online spaces.

E'zazi (Aramesh 2012) notes that imposing limitation on women in public space is a form of violence which justifies and reproduces violence against women in other sectors of society. Hence, reflecting on the main argument of the authorities for introducing women-only parks as venues to protect and reproduce religious values, Maryam, the young sociologist, sums up with the concluding remarks:

> With all due respect to those women who may like the women-only parks, I personally would never go to one. I would rather remove my *hijab* in public, and I practice that as a form of resistance whenever I get the slightest chance. A group of my friends and I started practicing it as "Stealthy Freedom"[3] long before it appeared as a trend in social media. This way I show my discontent about the compulsory *hijab* and gender segregation and all institutions and practices related to them. It's a kind of active articulation of disobedience and resistance. I don't like to follow what *they* have decided is best *for me* and I won't be thankful for something that I don't like or have never asked for. The parks may work for a small proportion of women of a certain age and background; to generalize their needs to all women is wrong.

Notes

1 One of the means of justifying and achieving the practice of power over the masses in post-revolutionary Iran was through the interpretation of the Quranic texts as, for instance, in IV: 59, to "Obey Allah, obey the Apostle and those in authority from among you." In the eyes of the ruling elite, "those in authority" are religious leaders

(*'ulama*) who are considered the true and righteous inheritors of the Prophet. Arjomand (1988: 12) argues that God does not use political authority; rather, he uses lordship over the universe. He is "not directly involved in mundane political events nor in the explicit source of political authority." However, the doctrine of acquisition (*kasb*) suggests that the notion of sovereignty as exercised by human beings is acquired and contingent upon the sovereignty of Allah. In practice, this idea led to the formulation of the *vilayat-i faqih* (authority of the jurisconsult). This doctrine flourished, and was both practiced by Ayatollah Khomeini and crystallized in the Islamic Republic of Iran. This view is not limited to any specific nation state, but aims at unifying Muslims worldwide in order "to create a government of universal justice in the world" (Khomeini, 1979: 66). For more in depth discussions see: Arjmand, 2008.

2 In its Bourdieuian sense, which is a setting in which agents and their social positions are located. The position of each particular agent in the field is a result of interaction between the specific rules of the field, agent's habitus and agent's (social, economic and cultural) capital (Bourdieu 2010).

3 Referring to *My Stealthy Freedom*, a virtual movement launched on Facebook in 2014 by an Iranian exile journalist Masih Alinejad which invites Iranian women to post pictures of themselves in public without a *hijab*. The movement received international attention and was awarded the Geneva Summit for Human Rights and Democracy Award (Dehghan 2014). The movement met the resentment of the Iranian religious and political authorities. Kazem Sadighi in his Friday prayer sermon in Tehran, criticized "corrupt messages circulating on the internet aiming to destroy Iranian families" (Nasseri 2014). A pro-government conservative news agency *Rajanews* called the movement "an obvious insult against Islam and religious establishment" (News 2014).

References

Abedi, Mehdi, and Micael Fischer. 2006. "An Iranian village boyhood." In *Struggle and survival in the modern Middle East*, edited by Edmund Burke and David N. Yaghoubian, 320–35, Berkeley: University of California Press.

Abu-Lughod, Janet L. 1971. *Cairo: 1001 years of the city victorious*. Princeton: Princeton University Press.

Abu-Lughod, Janet L. 1980. *Rabat: urban apartheid in Morocco*. Princeton: Princeton University Press.

AKA Iran. 2015. "Woman-only sport clubs." AKA Iran. Accessed 2015/08/15. http://www.akairan.com/sport/where-sport/11013.html.

Albardonedo Freire, Antonio José. 2002. *El urbanismo de Sevilla durante el reinado de Felipe II*. Sevilla: Guadalquivir Ediciones.

Allen, Terry. 1988. "Notes on Bust." *Iran* 26:55–68.

Amir-Ebrahimi, Masserat. 2006. "Conquering enclosed public spaces." *Cities* 23 (6):455–61.

Anton, Lorena. 2008. "Abortion and the making of the socilaist mother during communist Romenia." In *(M)Othering the nation: constructing and resisting national allegories through the maternal body*, edited by Lisa Bernstein, 49–61. Newcastle: Cambridge Scholars.

Aramesh, Hamideh. 2012. "From state violence to gender segragation in public spaces." *The Feminist School*. Accessed 2015/09/12. http://www.feministschool.com/spip.php?page=print&id_article=7202.

Aramesh, Sarah. 2010. *Raha tar az budan (Freer than being)*. Accessed 2015/06/05. http://rahatarazboodan.persianblog.ir/post/144/.

Arefi, Mahyar. 2014. *Deconstructing placemaking: needs, opportunities, and assets*. New York: Routledge.

Arendt, Hannah. 1998. *The human condition*. Chicago: University of Chicago Press.

Arjmand, Reza. 2008. *Inscription on stone: Islam, state and education in Iran and Turkey*. Stockholm: Stockholm University. Diss Stockholm Stockholms Universitet.

Arjmand, Reza and Maryam Ziari. 2016. "Why would I make things harder for myself: sexual concealment among Iranian young women." *Interdisciplinary Journal of Women's Study* (forthcoming).

Azad Armaki, Taqi. 2006. *Patuq and Iranian modernity*. Tehran: Loh-i Fekr.

Bamdad. 1979. "New regulations for barber guild." *Bamdad*.

Banerjee, Tridib, and Anastasia Loukaitou-Sideris. 2011. *Companion to urban design*. New York: Routledge.

Bartholomae, Christian. 1904. *Altiranisches Wörterbuch*. Strassburg: Verlag von Karl Trubner.

Bayat, Asef. 2010. "Tehran: paradox city." *New Left Review* 66 (November–December).

Benjamin, Samuel Greene Wheeler. 1887. *Persia and the Persians*. Boston: Ticknor and Company.

Bennison, Amira K., and Alison L. Gascoigne. 2007. *Cities in the pre-modern Islamic world: the urban impact of religion, state and society*. London: Routledge.

Bentley, Ian, Allen Alcock, Paul Murrain, Sue McGlynn, and Graham Smith. 1985. *Responsive environments: a manual for designers*. London: Architectural Press.

Bernstein, Lisa. 2008. *(M)Othering the nation: constructing and resisting national allegories through the maternal body*. Newcastle: Cambridge Scholars.

Bhabha, Homi K. 1994. *The location of culture*. London: Routledge.

Billig, Miriam. 2005. "Sense of place in the neighborhood, in locations of urban revitalization." *GeoJournal* 64:117–30. doi: 10.1007/s10708-005-4094-z.

Blookbashi, Ali. 2007. "Tawgh." In *The greater encyclopedia of Islam*, edited by Mohammad Kazem Bojnoordi, 372–5. Tehran: The Greater Encyclopedia of Islam.

Bosworth, Clifford Edmund. 2007. *Historic cities of the Islamic world*. Leiden: Brill.

Bourdieu, Pierre. 1984. *Outline of a theory of practice*. Cambridge: Cambridge University Press.

Bourdieu, Pierre. 1989. "Social space and symbolic power." *Sociological Theory* 7 (1. Spring):14–25.

Bourdieu, Pierre. 1990. *The logic of practice*. Cambridge: Polity.

Bourdieu, Pierre. 1991. "Physischer, sozialer and angeeigneter physischer Raum." In *Stadt Räum*, edited by M. Wentz, 25–34. Frankfurt: Campus.

Bourdieu, Pierre. 1996. *Physical space, social space and habitus*. Oslo: Institutt for sociologi og samfunnsgegrafi.

Bourdieu, Pierre. 1999. *The weight of the world: social suffering in contemporary society*. Oxford: Polity.

Bourdieu, Pierre. 2010. *Distinction: a social critique of the judgement of taste*. London: Routledge.

Bourdieu, Pierre, and Jean Claude Passeron. 1990. *Reproduction in education, society, and culture*. London: Sage.

Boyce, Mary. 2000. *Zoroastrians: their religious beliefs and practices*. London: Routledge.

Brownmiller, Susan. 1975. *Against our will: men, women and rape*. New York: Simon and Schuster.

Bulloch, John, and Harvey Morris. 1989. *The Gulf War: its origins, history and consequences*. London: Methuen.

Butler, Judith. 1999. *Gender trouble: feminism and the subversion of identity*. New York: Routledge.

Carmona, Matthew. 2003. *Public places-urban spaces: the dimensions of urban design*. Oxford: Architectural Press.

Carmona, Matthew, and Steven Tiesdell. 2007. *Urban design reader*. Oxford: Architectural Press.

Carr, Stephen. 1992. *Public space*. Cambridge: Cambridge University Press.

Carter, Jeremy G., Gina Cavan, Angela Connelly, Simon Guy, John Handley, and Aleksandra Kazmierczak. 2015. "Climate change and the city: building capacity for urban adaptation." *Progress in Planning* 95 (January 2015):1–66. doi: 10.1016/j. progress.2013.08.001.

Castells, Manuel. 1977. *The urban question: a Marxist approach*. London: Edward Arnold.

Central Intelligence Agency (CIA). 2015. "Iran." *CIA world factbook*. Accessed 2015/09/09. https://www.cia.gov/library/publications/the-world-factbook/geos/ir.html.

Chardin, John. 1811. *Journal du voyage du chevalier Chardin en Perse & aux Indes Orientales, par la Mer Noire & par la Colchide qui contient le voyage de Parie áa Ispahan*. A Londres: Chez Moses.

Clifford, Susan, and Angela King. 1993. *Local distinctiveness: place, particularity and identity*. London: Common Ground.

Colomina, Beatriz. 1992. *Sexuality and space*. New York: Princeton Architectural Press.

Connell, Raewyn. 2005. *Masculinities*. Berkeley: University of California Press.

Cook, M. 2012. "al-Nahy 'an al-Munkar." *Encyclopaedia of Islam*, second edition, edited by P. Bearman, Th. Bianquis, C.E. Bosworth, E. van Donzel, W.P. Heinrichs. Brill Online. Accessed 2015/03/12. http://referenceworks.brillonline.com/entries/encyclopaedia-of-islam-2/al-nahy-an-al-munkar-COM_1437.

Cooper, Rachel, Graeme Evans, and Christopher Boyko. 2009. *Designing sustainable cities*. Chichester: Wiley-Blackwell.

Cranz, Galen. 1980. "Women in urban parks." *Journal of Women in Culture and Society* 5 (3):79–95.

Davies, Llewelyn. 2000. *Urban design compendium*. London: English Partnerships/ Housing Corporation.

de Certeau, Michel. 1984. *The practice of everyday life*. Berkeley: University of California Press.

Dehghan, Saeed Kamali. 2014. "Iranian women post pictures of themselves without hijabs on Facebook." *The Guardian*, 12 May 2014.

DETER. 2000. *By design: urban design in the planning system – towards better practice*. London: DETER.

Deutsche Welle. 2015. *Air pollution claims life of one person in every other hour*. Berlin: Deutsche Welle.

Djamalzadeh, Mohammad Ali. 1985. Andarun. In *Encyclopædia Iranica* edited by Ehsan Yarshater, vol. 2, fasc.1, 11. New York: Columbia University.

Doan, Petra L. 2010. "Gendered Space." In *Encyclopedia of urban studies*, edited by Ray Hutchison, 298–302. Thousand Oaks: SAGE.

Duncan, Nancy. 1996. "Renegotiating gender and sexuality in public and private spaces." In *Body space destabilizing geographies of gender and sexuality*, edited by Nancy Duncan, 127–45. London: Routledge.

Eickelman, Dale F. 1981. *The Middle East: an anthropological approach*. Englewood Cliffs: Prentice-Hall.

Eilers, W. 1974. "Bagh." In *Encyclopaedia Iranica*, edited by Ehsan Yarshater, Vol. III, Fasc. 4, pp. 392–399. New York: Columbia University.

Eisenhauer, B.W., R. S. Krannich, and D. J. Blahna. 2000. "Attachment to special places on public lands: an analysis of activities, reason for attachments and community connections." *Society and Natural Resources* 12:421–41.

Eliade, Mircea. 1961. *Myths, dreams, and mysteries: the encounter between contemporary faiths and archaic realities*. New York: Harper.

Etemad-ul-Saltaneh, Mohammad Hossein Khan. 1999. *Merat-ul boldan (The mirror of the cities)*. Tehran: Tehran University Press.

Fakour, Mehrdad. 2000. "Garden." In *Encyclopædia Iranica*, edited by Ehsan Yarshater, vol. 10, fasc.3, 297–313. New York: Columbia University.

Fars News Agency. 2004. "Five parks in Tehran to be equipped for women." *Fars News.* Accessed 2015/08/12. http://farsnews.com/newstext.php?nn=8305200049.

Fars News Agency. 2005. "New parks for mothers in Teharn." *Fars News.* Accessed 2015/08/12. http://farsnews.com/newstext.php?nn=8409150328.

Fisher, Helen E. 2008. *The brain in love.* TED Talks.

Foucault, Michel. 1977. *Discipline and punish: the birth of the prison.* New York: Pantheon Books.

Foucault, Michel. 1979. *Les Machines à guérir: aux origines de l'hôpital moderne.* Bruxelles: Mardaga.

Foucault, Michel 1980. "The eye of power." In *Power/knowledge: selected interviews and other writings 1972–1977 by Michel Foucault,* edited by C. Gordon, 146–65. Sussex: Harvester Press.

Franck, Karen A., and Lynn Paxon. 1989. "Women and urban public space: research, design and policy issues." In *Public places and spaces,* edited by Irwin Altman and Ervin H. Zube, 120–44. New York: Plenum Press.

Fyfe, Nikolas, and Jan Bannister. 1998. "The eyes upon the street: closed-circuit television surveillance and the city." In *Images of the street: representation, experience and control in public space,* edited by Nikolas Fyfe, 254–67. London: Routledge.

Gal, Susan. 2005. "Language ideologies compared: metaphors and circulations of public and private." *Journal of Linguistic Anthropology* 15 (1):23–37. doi: 10.1525/jlin.2005.15.1.23.

Ganji, Neda. 2011. "Stop: no man is allowed here." *Donya-yi Eghtesad.*

Geertz, Clifford. 1973. "Thick description: toward an interprettive theory of culture." In *The interpretation of cultures: selected essays,* edited by Clifford Geertz, 3–30. London: Fontana.

Gehl, Jan. 1989. "A changing street life in a changing society." *Places* (Fall 1989):8–17.

Gehl, Jan. 1996. *Life between buildings: using public space.* Copenhagen: Arkitektens Forlag.

Gehl, Jan. 2011. *Cities for people.* London: Pidgeon Digital.

Giddens, Anthony. 2009. *Sociology.* Cambridge: Polity.

Ginsburg, Faye, and Rayna Rapp. 1991. "The politics of reproduction." *Annual Review of Anthropology* 20:311–43.

Glenn, H. Patrick. 2007. *Legal traditions of the world: sustainable diversity in law.* 3rd ed. Oxford: Oxford University Press.

Göle, Nilüfer. 1996. *The forbidden modern: civilization and veiling.* Ann Arbor: University of Michigan Press.

Golombek, Lisa. 2000. "Garden: Islamic period." In *Encyclopaedia Iranica,* edited by Ehsan Yarshater, 298–305. New York: Encyclopædia Iranica Foundation.

Grosz, Elizabeth. 1992. "Bodies-cities." In *Sexuality and space,* edited by Beatriz Colomina, 241–54. New York: Princeton Architectural Press.

Hafiz, Shams-ul-din Muhammad. 1999. *Divan-i Hafiz.* [Online]. Interpreted by Shahriar Shahriari. Accessed 2016/02/03. http://www.hafizonlove.com.

Hansen, Thomas Blom. 2008. "Race, security and special anxities." In *Gendering urban space in the Middle East, South Asia, and Africa,* edited by Martina Rieker and Kamran Asdar Ali, 131–4. New York: Palgrave Macmillan.

al-Harithy, Howayda. 1994. "Female patronage of Mamluk architecture in Cairo." *Harvard Middle Eastern and Islamic Review,* 1, (2):152–74.

Herdt, Gilbert H. 1994. *Third sex, third gender: beyond sexual dimorphism in culture and history.* New York: Zone Books.

Hildebrand, Frey 1999. *Designing the city: towards a more sustainable urban form.* London: Spon.

Hillier, Bill. 1996. *Space is the machine.* New York: Cambridge University Press.

Hillier, Bill, and Julienne Hanson. 1984. *The social logic of space.* Cambridge: Cambridge University Press.

Hillier, Jean, and Emma Rooksby. 2002. *Habitus: a sense of place.* Aldershot: Ashgate.

Imrooz. 2011. "Women-only post banks in Tehran." *Imrooz.* Accessed 2015/08/14. http://archives.emruznews.com/2013/04/post-16655.php.

Iranian Bureau of Statistics. 2011. *Census data.* Tehran: Iranian Statistics Bureau.

IRIB, Islamic Republic of Iran's Broadcasting Service 2013. "Tahmineh Milani opposing the women-only parks." In *Aparat*, edited by IRIB.

Isfahan Municipality. 2015. "Isfahan Municipality's Organization for Parks and Green Spaces." Isfahan Municipality. Accessed 2015/08/14. http://khadamate-shahri.blogfa.com/post-7.aspx.

Isfahan Municipality's Digital Portal. 2015. "Organization for Parks and Green Spaces." Isfahan Municipality. Accessed 2015/08/14. http://isfahan.ir/ShowPage.aspx?page_=form&order=show&lang=1&sub=39&PageId=2597&codeV=1&tempname=moarefiparkha.

Jacobs, Jane. 1969. *Death and life of great American cities.* New York: Vintage.

Jayyusi, Salma Khadra, Renata Holod, Attilio Petruccioli, and Andrâe Raymond. 2008. *The city in the Islamic world. 2 vols, Handbook of Oriental studies Section 1, The Near and Middle East.* Leiden: Brill.

Johansen, Baber 1979. "Eigentum, familie und obrigkeit im Hanafitischen Strafrecht." *Die welt des Islams*, 19(1):1–73.

Jomhoori-ye Islami. 2010. "Boys and girls have got out of control in parks and streets." *Jomhoori-ye Islami.*

Jorgensen, Bradley S., and Richard C. Stedman. 2006. "A comparative analysis of predictors of sense of place dimensions: attachment to, dependence on, and identification with lakeshore properties." *Journal of Environmental Management* 79:316–27.

Kakar, Palwasha L., and Brandy Bauer. 2003. "Space: domestic: Iran and Afghanistan." In *Encyclopedia of women and Islamic cultures*, general editor Suad Joseph. Brill. Accessed 2015/11/03. http://referenceworks.brillonline.com/entries/encyclopedia-of-women-and-islamic-cultures/space-domestic-iran-and-afghanistan-EWICCOM_0280b

Kamalipour, Yahya. 2010. *Media, power and politics in the digital age: the 2009 presidential election uprising in Iran.* Lanham: Rowman and Littlefield Publishers.

Kandiyoti, Deniz. 1996. *Gendering the Middle East: emerging perspectives. 1st ed., Gender, culture, and politics in the Middle East.* New York: Syracuse University Press.

Karimi, Pamela. 2003. "Cities: urban built environments: Iran." In *Encyclopedia of women and Islamic cultures*, edited by Suad Joseph and Afsaneh Najmabadi, 28–9. Leiden: Brill.

Karsh, Efraim. 2002. *The Iran–Iraq War 1980–1988.* Oxford: Osprey.

Kawsari, Samaneh. 2008. "A study on comments and recommendations of the Behisht Madaran Women-only parks users." Tehran: Tehran Municipality Office of Social and Cultural Research.

Kayhan. 1981. "The presence of sisters in the uban parks." *Kayhan.*

Kayhan Havaie. 1979. "Physical activities of iranian women." *Kayhan havaie.*

Khamenie, Ayatollah. 2014. "Supreme leader's speech in meeting with outstanding women." Accessed 2014/25/04. http://www.leader.ir/langs/en/index.php?p=contentShow&id=11703.

Konijnendijk, C. 2005. *Urban forests and trees: a reference book.* Berlin: Springer.

Koopayi, Abuzar Majlesi, Mojtaba Ansari, Mohammad Reza Bemanian, and Farhar Fakhar Tehrani. 2013. "Amin al-Doleh park: characterstics of Tehran's first urban park." *Bagh-i Nazar* 25 (2):3–19.

Krier, Leon. 1990. "Urban components." In *New classicism: omnibus,* edited by A. Papadakis and H. Watson, 96–211. London: Academy Editions.

Kuchelmeister, Guido. 1998. *Urban forestry: a development tool.* New York: United Nations, Tree City Initiative.

Laclau, Ernesto. 1982. "Diskurs, hegemonie und politik." *Neue soziale bewegungen und Marxismus,* edited by W.F. Haug and W. Elfferding, *Argument,* (Special issue) 17:807–13.

Lahiji, Shahla, the head of *Intishàràt-i rawshangaràn va mutale=at-i zanàn.* 2006. Interview by Khadijah Talattof, 2006/01/18, Tehran (authors' material).

Lambert, Léopold. 2013. *Foucault.* New York: The Funambulist.

Landman, Karina, and Martin Schönteich. 2002. "Urban fortress: gated communities as a reaction to crime." *African Security Review* 11 (4):71–85.

Lawson, Bryan. 2001. *The language of space.* Boston: Architectural Press.

Lefebvre, Henri. 1991. *The production of space.* Oxford: Blackwell.

Lefebvre, Henri, and Kanishka Goonewardena. 2008. *Space difference, everyday life: reading Henri Lefebvre.* London: Routledge.

Libson, Gideon. 1997. "On the development of custom as a source of law in Islamic Law." *Islamic Law and Society* 4 (2):131–55.

Libson, G., and F.H. Stewart. 2012. "'Urf." In *Encyclopaedia of Islam,* second edition, edited by: P. Bearman, Th. Bianquis, C.E. Bosworth, E. van Donzel, W.P. Heinrichs. Brill Online. Accessed 2015/02/22. http://referenceworks.brillonline.com/entries/encyclopaedia-of-islam-2/urf-COM_1298.

Lico, Gerard Rey A. 2001. "Architecture and sexuality: the politics of gendered space." *Humanities Diliman* 2 (1):30–44.

Loukaitou-Sideris, Anastasia, and Tridib Banerjee. 1998. *Urban design downtown: poetics and politics of form.* Berkeley: University of California Press.

Low, Setha M., Dana Taplin, and Suzanne Scheld. 2005. *Rethinking urban parks: public space and cultural diversity.* Austin: University of Texas Press.

Lynch, Kevin. 1960. *The image of the city.* Cambridge: MIT Press.

McDowell, Lisa. 1983. "Towards an understanding of the gender division of urban space." *Society and Space* 1:59–72.

McGlynn, Sue, Graham Smith, Paul Murrain, and Alan Alcock. 2013. *Responsive environments: A manual for designers.* London: Routledge.

Madanipour, Ali. 2003. *Public and private spaces of the city.* London: Routledge.

Mahdavi Far, Maziar. 2015. "Vitamin D deficiency epidemic affecting Iranian life." *Radio Zamaneh.*

MAI. 2011. *Census data.* Tehran: Iranian Bureau of Statistics (MAI).

Manucipality of Tehran. 2015. "About Tehran parks and green space organization." Manucipality of Tehran. Accessed 2015/08/15. http://parks.tehran.ir/Default.aspx?tabid=178.

Massey, Doreen 1992. "Politics and space/time." *New Left Review* 196:65–84.

Mbembe, Achile. 2003. "Necropolitics." *Public Culture* 15 (1):11–40.

Miranne, Kristine B., and Alma H. Young. 2000. *Gendering the city: women, boundaries and visions of urban life.* Lanham: Rowman and Littlefield.

Montgomery, John. 1998. "Making a city: urbanity, vitality and urban design." *Journal of Urban Design* 3 (1):93–116. doi: 10.1080/13574809808724418.

Morier, James Justinian. 1835. *The adventures of Hajji Baba of Ispahan.* Paris: Baudry's European Library.

Mottahedeh, Roy. 2004. *The mantle of the prophet: religion and politics in Iran.* New York: Pantheon Books.

Muqaddasi, Muḥammad ibn Ahmad. 2014. *Ahsan al-taqasim fi ma'rifat al-aqalim.* Leiden: Brill.

Mustawfi, Hamdallah. 1962. *My memoire or the social and adminstrative history of the Qajar era.* Tehran: Zavvar.

Najafi, Abolhasan. 2006. *Persian urban dictionary.* vol. 1. Tehran: Niloofar.

Najmabadi, Afsaneh. 1991. "Hazards of modernity and morality: women, state and ideology in contemporary Iran." In *Women, Islam and the state*, edited by Deniz Kandiyoti, 48–76. Basingstoke: Macmillan.

Nasseri, Ladan. 2014. "Iran conservatives hit back at Facebook campaign on headscarves." *Bloomberg.*

Oc, Taner, and Steven Tiesdell. 1997. "The death and life of city centres." In *Safer city centres: reviving the public realm*, edited by Taner Oc and Steven Tiesdell, 1–21. London: Chapman.

Paidar, Parvin. 1995. *Women and the political process in twentieth-century Iran.* Cambridge: Cambridge University Press.

Pinder-Wilson, Ralph. 1976. "The Persian garden: bagh and chahar bagh." In *The Islamic garden*, edited by Elisabeth B. MacDougall and Richard Ettinghausen, 69–85. Washington: Dumbarton Oaks, Trustees for Harvard University.

Pinder-Wilson, Ralph. 1985. *Studies in Islamic Art.* London: Pindar Press.

Prigge, Walter. 2008. "Reading the urban revolution: space and representation." In *Space difference, everyday life: reading Henri Lefebvre*, edited by Henri Lefebvre and Kanishka Goonewardena, 46–61. London: Routledge.

Radio Zamaneh. 2013. "Iran official says persons with disabilities need accessible public infrastructure." Global Accessibility News.

Rajanews 2014. "Stealthy freedom and insulation against Islam." *RajaNews.*

Ranchod-Nilsson, Sita, and Mary Ann Tetreault. 2000. "Gender and nationalism: moving beyond fragmented conversations." In *Women, states and nationalism: at home in the nation?*, edited by Sita Ranchod-Nilsson and Mary Ann Tetreault, 1–17. New York: Routledge.

Rapoport, Amos. 1969. *House form and culture.* Englewood Cliffs: Prentice-Hall.

Reddy, Gayatri. 2005. *With respect to sex: negotiating hijra identity in South India.* Chicago: University of Chicago Press.

Rendell, Jane, Barbara Penner, and Iain Borden. 2000. *Gender space architecture: an interdisciplinary introduction.* New York: E & FN Spon.

Rieker, Martina, and Kamran Asdar Ali. 2008. "Gendering urban space." In *Gendering urban space in the Middle East, South Asia and Africa*, edited by Martina Rieker and Kamran Asdar Ali, 1–15. New York: Palgrave Macmillan.

Rizvi, Kishwar. 2000. "Gendered Patronage: women and benevolence during the early Safavid Empire." In *Women, patronage, and self-representation in Islamic Societies*, edited by Ruggles Fairchild, 123–53. Albany: State University of New York Press.

Rosaldo, Michelle. 1980. "The use and abuse of athropology: reflections on feminism and cross-cultural understanding." *Signs* 5 (3):389–417.

Ruggles, D. Fairchild. 2014. "Gardens." In *Encyclopaedia of Islam, THREE*, edited by Kate Fleet, Gudrun Krämer, Denis Matringe, John Nawas, Everett Rowson. Brill Online. Accessed 2015/03/26. http://referenceworks.brillonline.com/entries/encyclopaedia-of-islam-3/gardens-COM_27373.

Sanders, Joel. 1996. *Stud: architectures of masculinity*. New York: Princeton Architectural Press.

Sedghi, Hamideh. 2007. *Women and politics in Iran: veiling, unveiling and reveiling*. Cambridge: Cambridge University Press.

Serat News. 2009. "Iranian National is gender divided". *Serat*.

Sharistanian, Janet. 1987. *Beyond the public/domestic dichotomy: contemporary perspectives on women's public lives*. New York: Greenwood Press.

Sheller, Mimi, and John Urry. 2000. "The city and the car." *International Journal of Urban and Regional Research* 24:737–57.

Sircus, J. 2001. "Invented places." *Prospect* 81 (September/October):30–5.

Smith, Neil. 2010. *Uneven development: nature, capital, and the production of space*. London: Verso.

Solheim, Jorun. 2001. *Den öppna kroppen: om könssymbolik i modern kultur*. Göteborg: Daidalos.

Spain, Daphne. 1992. *Gendered spaces*. Chapel Hill: University of North Carolina Press.

Spirn, Anne Whiston. 1984. *The granite garden: urban nature and human design*. New York: Basic Books.

Spivak, Gayatri Chakravorty. 2003. *Death of a discipline*. New York: Columbia University Press.

Steiner, Frederick R., and Kent S. Butler. 2007. *Planning and urban design standards*. Hoboken: John Wiley.

Stoler, Ann Laura. 2001. "Matters of intimacy as matters of state: a response." *Journal of American History* 88 (3):893–7.

Stronach, David. 1976. *Pasargadae: a report on the excavations conducted by the British Institute of Persian Studies*. Oxford: Clarendon Press.

Sultanzadeh, Hossein. 2013. "From bagh to park." *Name-ye Insan Shinasi* 1 (4):91–113.

Tabor, Philip. 2001. "I am a videocam." In *The unknown city: contesting architecture and social space*, edited by Iain Borden, 122–37. Cambridge: MIT Press.

Taylor, Affrica. 1998. "Lesbian space: more than one imagined territory." In *New frontiers of space, bodies and gender*, edited by Rosa Ainley, 129–41. London: Routledge.

Tehran Municipality. 2003. "Objectives and characteristics of Bihisht Madaran Park." Accessed 2015/01/05. http://region3.tehran.ir/Default.aspx?tabid=861.

Tehran Municipality. 2016. "Tehran: Statistics and data." Accessed 2016/01/22. http://services1.tehran.ir/Default.aspx?tabid=233&ctl=LayoutView&SId=292&YId=1392&mid=679.

Tehran Municipality Office of Social and Cultural Research. 2011. *A study on the satisfaction of the Tehrani citizens on green spaces and parks*. Tehran: Tehran Manucipality Office of Social and Cultural Resaerch.

Tehran Municipality's Statistics and Information Bureau. 2015. "Tehran Municipality's indicators: environemnt and green spaces." Tehran Municipality's Statistic and Information Bureau Accessed 2015/08/16. http://services1.tehran.ir/Default.aspx?tabid=233&ctl=LayoutView&SId=292&YId=1392&mid=679.

Thomas, Helen, and Jamilah Ahmed. 2004. *Cultural bodies: ethnography and theory*. Oxford: Blackwell.

Thys-enocak, Lucienne. 2008. "The gendered city." In *The city in the Islamic world,* edited by Salma Khadra Jayyusi, Renata Holod, Attilio Petruccioli and Andrâe Raymond, 877–93. Leiden: Brill.

Tonkiss, Fran. 2005. *Space, the city and social theory: social relations and urban forms.* Cambridge: Polity.

TPGSO, Tehran's Parks and Green Spaces Organization. 2015. "Bustan Pardis Banvan." *Organization of Tehran's Parks and Green Spaces.* Accessed 2015/08/12. http://parks.tehran.ir.

TPGSO, Tehran's Parks and Green Space Organization. 2016. "Women parks." Accessed 2016/01/27. http://parks.tehran.ir/Default.aspx?tabid=138.

Vista News Hub. 2015. "Women-only swimming pools in Tehran."

West, Paige, James Igoe, and Dan Brockington. 2006. "Parks and peoples: the social impact of protected areas." *Annual Review of Anthropology* 35:251–77. doi: 10.1146/annurev.anthro.35.081705.123308.

Will, Thomas E. 1894. *Parks and playgrounds.* New York: Arena.

Williams, Katie. 2000. "Does intensiying cities make them sustainable?" In *Achieving sustainable urban form*, edited by Katie Williams, Elizabeth Burton and M. Jenks, xii, 388. London: E. & FN Spon.

Women News and Analysis Agency. 2013. "Inactivity: a more serious health concern than cancer." *Women News and Analysis Agency.*

Index

Taylor & Francis eBooks

Helping you to choose the right eBooks for your Library

Add Routledge titles to your library's digital collection today. Taylor and Francis ebooks contains over 50,000 titles in the Humanities, Social Sciences, Behavioural Sciences, Built Environment and Law.

Choose from a range of subject packages or create your own!

Benefits for you
» Free MARC records
» COUNTER-compliant usage statistics
» Flexible purchase and pricing options
» All titles DRM-free.

Benefits for your user
» Off-site, anytime access via Athens or referring URL
» Print or copy pages or chapters
» Full content search
» Bookmark, highlight and annotate text
» Access to thousands of pages of quality research at the click of a button.

eCollections – Choose from over 30 subject eCollections, including:

Archaeology	Language Learning
Architecture	Law
Asian Studies	Literature
Business & Management	Media & Communication
Classical Studies	Middle East Studies
Construction	Music
Creative & Media Arts	Philosophy
Criminology & Criminal Justice	Planning
Economics	Politics
Education	Psychology & Mental Health
Energy	Religion
Engineering	Security
English Language & Linguistics	Social Work
Environment & Sustainability	Sociology
Geography	Sport
Health Studies	Theatre & Performance
History	Tourism, Hospitality & Events

For more information, pricing enquiries or to order a free trial, please contact your local sales team: www.tandfebooks.com/page/sales

Routledge
Taylor & Francis Group

The home of Routledge books

www.tandfebooks.com

Printed and bound by CPI Group (UK) Ltd, Croydon, CR0 4YY

22/10/2024

01777628-0020